ONE SIMPLE THING

ONE
SIMPLE
THING

A NEW LOOK *at the* SCIENCE
of YOGA *and* HOW IT CAN
TRANSFORM YOUR LIFE

EDDIE STERN

Foreword by Deepak Chopra

NORTH POINT PRESS
A division of Farrar, Straus and Giroux
New York

North Point Press
A division of Farrar, Straus and Giroux
175 Varick Street, New York 10014

Library of Congress Cataloging-in-Publication Data
Names: Stern, Eddie, 1967– author.
Title: One simple thing : a new look at the science of yoga and how it can
    transform your life / Eddie Stern ; foreword by Deepak Chopra.
Description: First edition. | New York : North Point Press, 2019. |
    Includes bibliographical references.
Identifiers: LCCN 2018035844 | ISBN 9780865478398 (hardcover)
Subjects: MESH: Yoga | Work of Breathing | Mind-Body Relations,
    Metaphysical
Classification: LCC RA781.7 | NLM QT 260.5.Y7 | DDC 613.7/046—dc23
LC record available at https://lccn.loc.gov/2018035844

Designed by Richard Oriolo

Our books may be purchased in bulk for promotional, educational,
or business use. Please contact your local bookseller or the Macmillan
Corporate and Premium Sales Department at 1-800-221-7945, extension 5442,
or by e-mail at MacmillanSpecialMarkets@macmillan.com.

www.fsgbooks.com
www.twitter.com/fsgbooks • www.facebook.com/fsgbooks

1 3 5 7 9 10 8 6 4 2

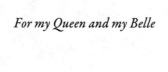

For my Queen and my Belle

CONTENTS

FOREWORD

BY DR. DEEPAK CHOPRA

IT SEEMS PECULIAR on the face of it that the mind is a problem, and even more peculiar that the mind is a problem for the mind. But the evidence of mental suffering is rife in modern society (one statistic reveals that Americans who are on a long-term regimen of antidepressants has doubled since 2010, and millions more are on long-term medication for anxiety). Any solution that might end mental suffering would be greeted with wild hope and relief—or so you would think.

It's possible for the mind to get so lost in itself that a person's very identity becomes confused, conflicted, and obscured. When Rumi asks, "Who am I in the midst of this thought traffic?" he speaks for every modern person. The sheer chaos of the mind is frightening, and finding an end to suffering by diving into oncoming thought traffic doesn't work.

Eddie Stern's insightful, wide-ranging, and at times eloquent book on yoga takes an optimistic view of how mental suffering can end. He doesn't focus only on the aspect of yoga as healing. There is ample treatment in these chapters of the underlying philosophy of yoga, its universality around the world, its potential for helping the poor and outcast who are gripped by the worst kind of suffering, including prison populations. But healing through "one simple thing," the regular, dedicated practice of yoga, is the essence of his message.

To accept healing is difficult, and mental healing the most difficult of all. Rumi confronted a mind filled with teeming, seemingly random thoughts, and our mental landscape today, distracted by video games, social media, and the Internet, would be totally foreign to medieval Persia or ancient India. Yet Rumi and every other fully conscious person who has waked up knows that people will spend a lifetime choosing to be in denial, afraid of their own impulses and desires, driven by those same impulses and desires, totally convinced that the darker aspects of the psyche must be suppressed, and deeply embedded in social conformity.

When William Blake walked through the streets of eighteenth-century London, the "marks of weakness, marks of woe" he saw in the passing crowd were the result of "mind-forg'd manacles," a haunting phrase that I have kept in the back of my mind for three decades. When the mind functions as both jailer and prisoner, finding an end to mental suffering seems incredibly difficult. Even motivating people to try is daunting.

Eddie Stern, with his long experience as a yoga teacher and lecturer, understands everything about motivating his students and prospective students. On a personal level, this may be the most persuasive part of this book, beginning with careful bridge-building from the long-ago India of the *rishis* to the modern secular world. Yoga currently rides a crest of popularity, but trends are fickle, and Stern knows that unless there is more than regular yoga class, unless there is a complete vision of yoga's potential, there is a real risk of yoga becoming a passing phenomenon.

The underlying vision rests on yoga as union, which means overcoming the divided self. Separation is the opposite of union, and the ultimate separation, which has affected all of us, is the mind in separation from its essential nature. You can approach the issue from many angles, as this book skillfully does. A kidney, heart, or lung cell is already unified in its natural state. Cells don't doubt their existence. They function holistically and offer us a model for life as a flow of energy and intelligence. So the connection between mind and biology forms a strong theme in Stern's teaching about yoga.

One can also focus on other distressing signs of separation, in troubled relationships, social discord, and all manner of self-destructive behavior, including addictions and preventable lifestyle disorders that people exacerbate rather than helping themselves to heal. Yet in the end, it is the self divided against itself that yoga fundamentally addresses. The mind would not be its own enemy except for the divided self; the body would not be abandoned as a thing to be ignored, shunned, or ashamed of (except for those gifted with superb and beautiful bodies, and even they must face the specter of time and aging).

One of the most important tenets of yoga is that the level of the problem isn't the level of the solution. As long as we remain inside the state of self-division, we are dominated by it. There are only three attitudes one can take to mental suffering: put up with it, fix it, or walk away from it. Unfortunately, all three are doomed to failure—and for the same reason. The mind that attempts to put up with suffering, fix it, or escape from it is the very mind that has been split by the state of separation. A fragmented mind is like Humpty Dumpty, whose fall is misunderstood. All the king's horses and all the king's men can't repair a broken egg, no doubt. But Humpty Dumpty can't put himself together again, which is the real problem.

Yoga solves this dilemma by asserting some home truths that are then carried out through the practice of yoga and meditation. I've already given the first one, that the level of the problem is never the level of the solution. Here are the other home truths, as I understand them:

The level of the solution is consciousness, which in its very nature is whole, complete, and undivided.

Consciousness, being the source of creation, is always present in its pure, whole form.

When the mind experiences its source in pure consciousness, solutions dawn, not through the effort to end suffering but through the state of wholeness itself—no outside agency, motivation, or thinking is required.

The body, brain, mind, and universe are different modes of consciousness. Each mode is self-regulating, and so is the whole. A cell has the capacity to keep itself alive and thriving, in a state of perfect balance. The same is true at every level of Nature.

When self-regulation fails, the underlying cause is loss of contact with wholeness. By experiencing pure consciousness, self-regulation is restored. This returns body, brain, mind, and universe to their unified state.

I know that using yoga to return the universe to its natural state seems unbelievable, and the claim is too vast to explore in a page or two. But when yoga returns us to our source and allows us to experience the awakened state, there is no alternative but for reality itself, which we dub the universe, to shift along with everything else.

For all its current popularity, yoga in India is entangled in mind-numbing intricacies, philosophical controversy, endless wrangling over the ancient texts, competing teachers and systems—on and on. I deeply admire Eddie Stern for bringing clarity and compassion to such an unholy mess—the future depends on such clarity and compassion. Taken altogether, this book is the gentlest and most accessible way to embrace yoga in all its potential. Insofar as humans have infinite potential—another home truth about consciousness—yoga unfolds a field of possibilities that would otherwise be closed to the divided self. Stern never lets us forget our untapped potential, which may be the core lesson of his teaching and his life.

ONE SIMPLE THING

INTRODUCTION

IN THE SPRING OF 2010, a researcher and physical therapist named Marshall Hagins came to visit me at my yoga school in SoHo, in New York City. He wanted to see if I would be interested in designing a yoga protocol for a scientific study to examine whether yoga could have a positive impact on pre-hypertensive conditions in African Americans. Twenty-two years earlier, I had bypassed college, opting instead for travel to India, and had spent the better part of the years since returning to India, studying, practicing, and reading everything on yoga that I could. I was quite comfortable with yoga, but I was clueless regarding even the basic tenets of science. Here was a very smart person sitting in front of me, thinking I knew something and asking for my help. So, with or without scientific knowledge, I of course said "Yes!" Little did I know that this meeting would shift my focus within yoga away from studying and memorizing ancient texts and toward investigating what makes yoga work so well.

Why was it that a person with back pain, another with hypertension, another with poor digestion, others searching for meaning in their lives, could all walk into the same yoga class, do the same basic thing, and walk out not only feeling better but feeling like what was troubling them, or the condition that they had, was improving? How, by doing one simple thing, one generalized yoga practice, were people able to reduce stress, ease body pains, improve cardiovascular function, reduce diabetes medication, feel happier, get angry less often, and improve their sleep and digestion? Somehow, given the opportunity, the body knew how to correct imbalances. And even more interesting, it was apparent that the yoga poses didn't even have to be done "well" or "right" for these positive effects to happen; whether someone is stiff or flexible, thin or heavy, sick or healthy, yoga seems to work. I had an inkling that it had something to do with the nervous system, but I was not sure what. So I began to read, to speak to doctors, and to investigate.

This book is largely the result of that investigation (that, along with thirty years of yoga practice). In my discussions with doctors, I learned about the Western presentation of the nervous system. I was already familiar with what the yogic texts had to say about the nervous system and could look for correlates, make conjectures, and then discuss those correlates with the doctors. I would always check to see if they felt that what I was saying sounded valid, and slowly the ideas presented in this book became refined. I looked into the amazing research that has been and is still being conducted by Dr. Stephen Porges, Dr. Shirley Telles, Dr. Bethany Kok, and many others. I refer to their work throughout this book to shine a light on how science understands yoga to be an effective practice for nervous system self-regulation.

Two burning questions were developing in my mind as I studied.

1. Where do consciousness and biology meet? Does consciousness reflect itself through our biological makeup, or is our biology actually consciousness manifesting itself? We are human beings, after all, and the yogis used the body (which is biological) to journey inward toward "deeper" levels of consciousness. Therefore, it would follow that the body and consciousness are somehow linked.

2. Is happiness a physiological experience? That is, by seeking to discover who we truly are through yoga practices that use the body, breath, and mind, can we find a deep, lasting inner peace and happiness that exist in our physiological makeup? It seemed to me that because we use the body in every aspect of yoga, happiness cannot be simply a mental construct, but is *integral to our physical makeup and can be*

found in the inner mechanisms of our bodies. Perhaps transcendence lies beyond the realm of the body, but what of simple happiness and the ease of knowing who you are? Happiness as a mental construct seems to be an impossibility. We can't hold a thought in our minds for more than a second or two. So how can we *mentally* hold on to happiness? Happiness perhaps exists somewhere deeper. In the Hindu thought systems, happiness is equated not with pleasure, but with meaning or purpose. It is not happiness we are after, but the experience of our own essential being. We are seeking ourselves. Is this experienced in our physiology? Or are the mind and body a continuum, so there is an integrated experience of being because there is no distinction between body and mind?

These questions are largely my jumping-off point. And, as it happens, the nervous system is an integral and key component in examining them. The ancient yogis taught that the science of yoga is not about the perfection of postures, but perfection of the body-mind-spirit relationship, so that one can understand the deepest mysteries of being. Their teachings, and the teachings of my South Indian guru, impelled my journey and my continuing fascination with these topics. What follows in this book are explorations into yogic ideas and scientific research on the underlying neurobiological mechanisms that help explain how and why yoga has such a positive overall impact on our bodies, our minds, and the world, and how we can find happiness, meaning, and purpose in it.

{ WHAT IS YOGA? }

IF THERE IS A SPIRITUAL practice that has been mocked, lampooned, and stereotyped in the West, it's yoga. And why not? Western yogis are easy to make fun of. With our top-knots, expensive leggings, chia seeds, smoothies, yoga mat bags over our shoulders, extended retreats, crystals, namaste-ing, om-ing, and sitting cross-legged everywhere, if you want to make fun of us, there is plenty of material to pick from. A couple hundred years ago, yogis in India were also mocked and denigrated, during the time of the British occupation and by the early travelers who had never seen anything like them before.[1] Accounts as early as the one in 1689 by John Ovington describe the "painful and unnatural postures" of the ash-smeared, philosophical mendicants known as fakirs, a name for the Persian ascetics who were lumped into the same category as the Hindu yogis. The armed and highly organized ascetic order of the Naga Sannyasis[2] presented a violent challenge to the hegemony of the East India Company, and from the mid-1700s to the early 1800s the Naga Sannyasis and Muslim fakirs staged uprisings and attacks against the East India Company in Bengal, which eventually led to big crackdowns on all ascetic organizations.[3]

The fact that Westerners lumped the Nagas, fakirs, and yogis of more gentle orders together into the same category of dangerous and violent ascetics is perhaps one reason that yoga fell out of vogue in India in the 1700s and 1800s. However, even if the yogis were considered to be dangerous, as well as filthy, lying scoundrels (and there are examples of this view even today), the practice and philosophical tenets of yoga somehow made it through this rocky period in India and found a resurgence with Sri Krishnamacharya,[4] Swami Sivananda,[5] and yoga's journey to the West in the late 1800s. As of 2014, with Prime Minister Narendra Modi, an avid promoter and practitioner of yoga, at the helm of India, the birthplace of yoga has certainly begun to pull its own weight in yoga again.[6] Modi's suggestion of an International Yoga Day to foster global harmony and inner and outer peace was sponsored by every country represented at the United Nations, and has helped India reclaim its place of primacy in the world of yoga. Although some may say that India never lost touch with yoga, in the late 1980s I spent a lot of time traveling from North to South India looking for yoga teachers, and found relatively few of them. In 1990 there were only two or three yoga schools in Mysore—at present, there are close to fifty. Mysore is now considered to be one of the yoga capitals of India, and it is largely due to the influence of Pattabhi Jois. The yoga landscape in India has changed dramatically.

Yoga arrived in America in the 1800s and has been largely assimilated into our culture. Though Americans studied yogic texts at the beginning—Ralph Waldo Emerson loved the *Bhagavad Gita*—few actually practiced yoga. However, in the short span of a little more than two hundred years, millions upon millions of people across all walks of life have begun to practice yoga—in 2017 an estimated 36 million in the United States alone did some form of yoga—and not just by those who are on a spiritual path. It is practiced by children in schools, the elderly in chairs, people who are incarcerated, those who suffer from PTSD, patients in hospitals, and folks who just have a lot of stress in their lives.[7] Yoga provides solace, free of discrimination.

And yet it's also important to acknowledge that in the United States, at present, we find ourselves in the midst of a very real clash of cultures. In the 1960s we had East meets West, and the hippie movement, as a generation of youth tried to free themselves from the shackles of wartime austerity and restrictive nuclear-family ideals. As I have watched the yoga scene grow over the past thirty years, it's now more like West gobbles up the

East, and the free-form embrace of spirituality has veered into a head-on collision with consumerism—exactly the opposite of what yoga was supposed to promise and deliver. India, especially under Prime Minister Modi, has begun to reclaim yoga as part of its cultural heritage, which indeed it is. But in the meantime, the West has adopted yoga as one of its own children, and yoga in the United States has adapted to life here in unusual ways, including the secularization of a contemplative practice.

It is hard for me to separate the ancient Indian—or Hindu—culture from yoga practice, and I am not sure that turning a contemplative, mystical practice into a completely secular fitness regime is a good idea. Once you remove the contemplative aspect of yoga from its practice, can it truly be called yoga anymore?[8] On the other hand, yoga has proven itself to be beyond religions, and beyond religious beliefs, and that is readily seen in the people from a variety of religions and the non-religious who practice yoga because it calms their mind, reduces stress, and makes them more internally clear. A pastor who practices with me uses the time when he does deep breathing at the end of his practice to contemplate his Sunday sermon; a rabbi uses his practice to find a quiet space that his spoken prayer does not give him. The Judeo-Christian traditions all have mystical branches, wherein a direct relationship with the divine is sought, but the mystical branches are often seen as fringe movements. The Eastern traditions made no distinction between the world and the sacred. Yoga, ritual, and the earth were all seen as one; they were mystical to the core. Today we often forget that there is a difference between religion and mysticism, dogma and contemplation. And that is precisely where yoga excels. It is easy-access mysticism. It is instantly contemplative, usually from the first time you lie down and rest deeply after practice.

While some elements of yoga are deeply entwined in the Hindu tradition, others are not. There are hints in the ancient texts that, as a practice, yoga transcends culture, time, place, and what we now call religion.[9] While yoga is indeed from India, and rooted in Hindu thought systems, yoga has proven itself to be extremely adaptable, and is practiced on every continent by people from varied backgrounds and different cultural perspectives. The remarkable thing about that is that of the millions who are practicing yoga with regularity, many have very similar results: we feel better, are more clear-headed, are healthier, and in many cases have a deeper sense of purpose. This is a hint to what the basis for the Hindu tradition, called the Eternal Way, or Sanatana Dharma, was before it was called Hinduism. Beyond deities or reincarnation, Hinduism is concerned with the idea that every being has an essential purpose, and that we should strive to live our lives in such a way for that purpose to be fulfilled. It is from this vantage point that I view yoga.

So many things in the world divide us, such as politics, religion, sports teams, and all of our personal opinions, ideas, and judgments. It is rare to find something that connects us. Yoga is one of those things, and it has the ability to help us transcend partisan distinctions because it has clarity of mind, compassion, empathy, kindness, love, and caring as its base— these are all mental states and emotions that transcend religion, distinction, and things that set us apart from each other. They are things that connect us, or remind us of our connectivity, and not the things that divide us. Of course, in the marketplace, we do not always see this reflected, but when it comes to the results that people experience from yoga, the benefits are largely the same. I find that extremely interesting, and it is one of the things that led me to ask, What is the underlying mechanism that makes yoga work for so many different people, almost regardless of the type of yoga that they practice?

THE WORD *YOGA*

The word *yoga* has several meanings. Among them are "union," "concentration," "a path," and "relation." The word itself comes from the verbal root *yuj*, which means "to yoke or join" and is why the word *yoga* is most commonly associated with the idea of union. The ancient Sanskrit grammarian Panini wrote that there were two ways of defining the word *yoga*, depending on usage. The first is *yujir yoge*, which describes the action of joining or yoking—for example, the joining of an ox to a cart. In the earliest teachings of the ancient Sanatana Dharma canon, called the Vedas, this was the sense in which the word *yoga* was used. But for yoga practice, which was classified as a spiritual discipline in later years during the Upanishadic age (800–500 B.C.E.), the correct derivation is from *yuj samadau*, which means, roughly, that yoga is a special type of concentration, called *samadhi*.[10] *Samadhi* means "absorption," and it is a natural tendency of the mind to become absorbed in things, whether thoughts, objects, work, ideas, a love interest, or goals. When it comes to absorbing the mind in spiritual pursuits, the mind is said to take on the form of that which we are contemplating, and eventually, that deep level of absorption leads to the insight and experience of our true nature. In the deepest level of samadhi, one gains knowledge of one's inner being, or self.

CONCENTRATION

Around twelve hundred to two thousand years ago, a sage (rishi) named Patanjali collected the existing teachings on yoga and systematized them in a form known as sutras. The sutra form of authorship means that the author did not create a system and write an original work, but compiled the teachings, practices, and methods that already existed, collecting or codifying them under one heading. There are six philosophical schools in Hinduism, and each has sutras that contain their teachings.[11] For yoga, the text is called *Patanjali Yoga Sutras*, and it contains 196 sutras. A sutra is a short sentence, the few words of which have a much larger meaning. Commentaries given on the short forms by other sages and saints fill in the details and elaborate on the finer points of what the sutra is actually saying, because more often than not, they are quite hard to understand at face value.

Patanjali explained in his *Yoga Sutras* that *samadhi*, the highest state of concentration, is a technical term for the mind's innate ability to become absorbed in its object of contemplation. Patanjali's is not the only presentation of yoga, but it is indeed one of the most complete. Other presentations of yoga that came after Patanjali have different end goals, but all of them have one thing in common, namely, the idea that in order to achieve your goals, you need to be able to focus your mind. Therefore, Patanjali defines *yoga* in the second sutra of his book as the ability to selectively eliminate all extemporaneous thoughts or movements that occur in the mind and to choose where you want your mind to be, or where you want to focus it.[12] As my Sanskrit teacher Vyaas Houston has said, the *Yoga Sutras* serve as a road map for inner consciousness. These short, concise aphorisms, which are packed with meaning, lead us through deeper and deeper levels of our mind, consciousness, and reality. Many of the teachings contained within the sutras—several of which will be discussed in this book—are amazingly relevant to us even today. Why is this so? It's because, I think, that the mind we have today is no different from the minds that people had two thousand, or even five thousand, years ago. We suffer, we struggle, we experience joy and desire, and we question, we investigate. The quest to know ourselves, to question who we are and what we are doing here is not new to us; it is in fact a part of us to question like this, and it is this impulse that drove people to create systems of yoga thousands of years ago, and is the same impulse that drives so many to practice it today.

The earliest commentary on the *Yoga Sutras* was written by the ancient sage Vyaas (a different Vyaas than my Sanskrit teacher). In his commentary, he discusses how the mind has five basic patterns, or states.[13] We can clearly see that these five patterns have not really changed at all in two thousand years. The first two states are not conducive to yoga practice, but the last three are. However, it is only the last two states that are conducive to samadhi, or complete absorption. The states are:

1. Restless

2. Stupefied

3. Distracted

4. One-pointed

5. Completely restrained

A person with a restless mind will never want to practice yoga, because he or she cannot remain focused for any length of time. The mind jumps from here to there, never staying fixed for even a moment, like having attention deficit disorder. I know plenty of people with ADD who are very productive and successful people, but they struggle to do yoga consistently, and sometimes find that meditation practices like Transcendental Meditation (TM) are easier for them. A person with a stupefied mind is obsessed with their problems, and ruminates, turns, and dwells on them. We've all had the experience of a problem, conflict, heartbreak, or disappointment becoming the only thing that we can think or talk about, sometimes to the point where our family or friends will want to shake us and yell "Get over it!" The stupefied mind has a hard time doing any type of contemplative practice, or doing anything at all, for that matter, except obsess about its own problems. Obsessive-compulsive disorder is an extreme example of a problem of the stupefied mind.

The distracted mind, as odd a description of a spiritual practitioner's mind as it may be, is the state of mind of most of us who come to learn yoga. We are able to concentrate for short periods of time, but then we revert back into distraction. This is a state of mind that almost all yoga practitioners are familiar with: we can stay focused for a bit, but then our mind wanders off. The act of catching the mind after it has wandered off, and returning it to the place of our choosing, is one of the basic activities that we are training ourselves to do in yoga practice, and this is doable even with a mind prone to distraction. This is because one of the hallmarks of the distracted mind is that it can be calm at one moment, and then restless at another moment. The state of change that occurs in distraction is also the state that teaches us how to begin harnessing the power of attention—we get the opportunity to work on catching the mind when it becomes restless. People with this type of mind know that they need to do yoga or meditation because they have experienced both calmness and distraction, and would like to strengthen their ability to be in a calmer, more relaxed state. That is why it is said that the mind of the person who comes to yoga is predominantly in this third state, the distracted state. If you identify yourself as a person whose mind easily gives in to distraction, then I have some good news: you're the perfect candidate for yoga!

The final two states of mind—one-pointed and completely restrained—are the states that samadhi can occur in. "We should bear in mind," said Swami Hariharananda, "that our mental weakness is only our inability to retain our good intentions fixed in the mind; but if the fluctuations of the mind are overcome, we shall be able to remain fixed in our good intentions and thus be endowed with mental power. As the calmness [of mind] would increase, that power shall also increase. The acme of such calmness is Samadhi."[14] I particularly like this quotation, because this idea is clear: yoga is not about screwing the mind into a fixed state of focus, or the body into a complicated pose; it is about calmness and filling the mind with a natural state of goodness. It is a natural, underlying characteristic that has been covered up by too much thinking. Sometimes when I sit and meditate, I don't do anything but sense or feel for that natural state of goodness within me. Like many people, I judge myself pretty harshly; I prefer criticism over compliments because I would rather improve myself to the point of perfection, and hearing what was good just gets in the way of what needs to be fixed. But not everything needs to be fixed; it's okay to sometimes just let things be. So when I sit and feel the natural goodness that is inside of me, a feeling of calm does indeed automatically come to me. It's soothing because, from this point of view, goodness is not something that we strive to be or become; it is something that is already there. We just have to allow it to be a little more present.

In the last two states of mind on Vyaas's list, the one-pointed and completely restrained states, the deepest experience of samadhi occurs, also known as the "state of yoga." In the one-pointed state of mind, you can rest your attention on any object that you choose to contemplate—whether it be your breath, a mantra, or something else—for as long as you wish. That is no easy feat. It is hard to keep the mind resting on one thing for literally even a few seconds. In the completely restrained, or arrested, state of mind, there are no thoughts, no fluctuations, and no object separate from yourself to hold your mind to. Subject and object cease to exist, leaving non-localized consciousness as your only experience. Everywhere you look, listen, hear, smell, or touch, there is only consciousness. In the deepest states of samadhi, there are no longer any objects; only the subject remains. It is called *vishesha*, or that which is left over, after all the changing objects of the world no longer color our experience. This is sometimes referred to as "unity consciousness."

The South Indian yoga master Sri K. Pattabhi Jois wrote that the word *yoga* has several meanings. Among them are "relation," "a means," "union," "knowledge," "matter," and "logic."[15] He was unique in defining the practice of yoga according to one of Patanjali's sutras, 2.26, which states that yoga is an *upaya*, a path.[16] What kind of a path is it? One that brings to an end the confusion of the mind, via a special type of mental discrimination that leads toward self-knowledge, a discrimination that allows us to distinguish awareness from the movies of our lives, thoughts and desires that are projected onto its screen. Yoga practice, therefore, is the means of liberation from conditioned thinking.

Jois writes:

> For now, let us say that the meaning of the word yoga is upaya, which means a path, or way which we follow or by means of which we can attain something. What then is the path we should follow? What or whom should we seek to attain? The mind should seek to attain what is best . . . the way of establishing the mind in the Self should be known as yoga.[17]

The idea of upaya is intricately linked with the sense of relation that Jois lists first in his definition of yoga. For, by doing the yoga practice, and the related contemplative practices, we gain an intimate relationship with our body, breath, mind, emotions, and sense of purpose. It is this intimacy with ourselves that leads toward self-confidence and comfort with who we are and what we are doing here. This naturally leads to the deeper and most important question that we eventually end up asking ourselves: Who am I beyond the idea that the sum total of who I am is my body, my emotions, thoughts, or memories? These are our main questions in life: Who am I? What am I doing here? My ninth-grade English teacher, Mrs. Jane Bendetson, posed these questions to our class as the most important questions that we could ever ask ourselves, and added to them, What should I do next? These questions are, in fact, the only thing I remember having learned in high school.

Yoga first and foremost is a practice. The yogis considered that we should practice yoga in the same way, or with the same importance, that we brush our teeth each day. Through practicing yoga asanas (postures) and breathing, as will be discussed in each chapter of this book, we clean our body internally, and strengthen the muscles, bones, internal organs, nervous system, mind, and emotions. A little bit of practice goes a long way; we don't have to practice for hours on end every day, to the point of exhaustion. All we have to do is make sure that doing a little bit of practice each day becomes a priority in our lives, and that we do so until practice becomes a habit, a regular part of our daily routine, or a part of the ritual that makes up the rhythm of our life. Any practice, whether spiritual, physical, or artistic, only begins to pay off when it is done with regularity and sincerity. One of Patanjali's most quoted sutras, 1.13, is on this very point:

Sa tu dirgha kala nairantarya satkara sevito
drdha bhumih.

||||||

We become grounded in practice when it is done
uninterruptedly, for a long time, with devotion.

Perhaps more important than the idea of discipline is what
discipline is creating. The neuroscientist and psychologist Rick
Hanson has written about this at length in his book *Hardwiring
Happiness*, where he describes the difference between mental
states and mental traits. We are often the victims of our mental
states: anger, jealousy, judgment, revenge, laziness, apathy, boredom, desire; and at times we act on these states, and identify
with them. But these states are transient; they come and go.
Still, they are liable to repeat themselves more often when we
act upon them. By doing a regular practice, we begin to create an
underlying mental trait of awareness, which is more dependable
and more open than the changing states. Through our practice
we develop within us a trait of awareness that is calm, has perspective, and can help us to pause so that we do not get swept
away by overwhelming emotions.

Developing strong mental traits, then, is the true goal of a
dedicated practice. Patanjali does not define practice as being
really great at doing yoga postures; he defines it as a means to
creating an underlying mental trait of awareness that leads
toward insight. The changing *states* are what Vyaas was referring to when he spoke about the distracted mind, and one
of the first things that yoga gives is the ability to observe the
changing states without getting lost in them. Many have expe-

rienced that after practicing yoga for even a short time, they get angry less often, or catch themselves before speaking without thinking of the repercussions of their words. This is because the underlying trait of awareness is starting to become as pronounced, if not more so, than the changing states.

SADHANA, THE MEANS

As with many of the yogic ideas, and many Sanskrit words, one word will lead to another word that fleshes out even more subtleties of meaning. Yoga practice has a special word associated with it, *sadhana*, which describes the techniques or practices that we use to move toward self-knowledge, awareness, or liberation. *Sadhana* is often translated as "spiritual practice," and the purpose behind spiritual practice, usually, is liberation from suffering—which is liberation from identification with everything that is other than awareness. Sadhanas are the means that we use to identify with a sense of awareness within, and to remove the coverings of confusion, narratives, and longings that prevent us from being who we truly are.

A. G. Mohan, an influential yoga teacher from Chennai, said a wonderful thing about the different layers of meaning and experience that in the Hindu tradition have been compared to the peeling of an onion. This analogy is very often used to describe the layers of spiritual practice: you keep peeling and peeling the layers of identification away until there is nothing left but consciousness. "But," Mohan points out, "who is the one who has peeled the onion? The one who peeled the onion does not disappear as well." Sadhana is the peeling of the onion;

the one who has peeled the onion is the impulse within us to know.

Sadhana is a commitment to making our spiritual goals, in particular, a priority, and making time for them. A spiritual goal can be:

* Practicing yoga
* Practicing meditation
* Practicing kindness, gratitude, or forgiveness
* Living a balanced life
* Keeping our minds calm and accepting
* Serving those in need
* Living a contemplative, thoughtful life
* Practicing patience
* Becoming a better listener

If we say we want any of these things but are not taking active steps to actually do them, then we can't really say that we want them. If I say that I want to be more meditative in my life, but I don't make the time to practice meditation every day, then perhaps I do not really want to be meditative. The things that we actually spend time doing are the things that we want, and sometimes the goals or ideas we have are not real—they are just ideas that sound pretty good to us. In sadhana, it's important to figure out, What is it that I really want? And if I do really want that thing, then I'll spend time doing it. It's as easy as that.

Don't worry about not doing things you don't really want to do. If you say that you want to meditate but you never do it, then you probably don't want to meditate. If you accept that you don't want to meditate, then you won't feel bad about not doing it, and you can cross it off your list of things that you think you want to do—stuff that other people do that sounds like a good idea but, when push comes to shove, is not for you. Then you can replace it with something that you really do want to do. Sometimes we do actually want to learn or practice something, but we find it hard to make time for it—if that is the case, then you need to learn to be more disciplined, and to put up with a little hardship. In Sanskrit, this is called *tapas*. That is where satisfaction, success, and even excellence come from: overcoming the obstacle of either getting started or finishing something to completion. Knowing what you want is *sadhya*, or the goal; the path we travel to get there, the upaya, is sadhana.

As Timothy Ferriss says in his book *Tribe of Mentors*, "Life punishes the vague wish and rewards the specific ask. After all, conscious thinking is largely asking and answering questions in your own head. If you want confusion and heartache, ask vague questions. If you want uncommon clarity and results, ask uncommonly clear questions." These following three words lay out the concrete plan, or road map, of spiritual practice:

1. *sadhya:* forming our goal

2. *sadhana:* our practice, which is the means of accomplishing it

3. *upaya:* sticking to the path

The goal that we choose does not necessarily need to be liberation. The goal could simply be to move our bodies for thirty minutes a day for health reasons; it could be to meditate for seven minutes a day to calm our minds; it could be to chant a mantra 108 times to express devotion. The goal we choose should be attainable; otherwise we will get discouraged. If you can pick a small, attainable goal and reach it, then little by little your goals can become more subtle. For example, a goal of getting less angry, or not getting annoyed by small things. This will begin to happen naturally when you establish yourself in a daily discipline.

There is another definition of *upaya* that I quite like, and that is the one that is used in Jyotish, or Vedic astrology. An upaya in astrology is a remedy that the astrologer gives to someone who has a *dosha*, or defect, somewhere in their chart, that is causing them trouble or creating an obstacle in their lives. The astrologer may suggest to them that they repeat a particular mantra, wear a certain color, feed a particular type of animal, all on a specific day of the week, for a certain period of time, in order to remove this defect. Such an upaya is a remedy as ritual to remove an obstacle. In yoga, the biggest obstacle we have is an undisciplined mind that is attached to thinking about stuff all the time, that is attached to our opinions, judgments, and ideas, which lead us to create false identifications: I'm a Democrat, I'm a Republican, I'm a vegan, I'm an Ashtanga yogi, I'm an Iyengar yogi, I'm a bad person, I'm a great person. All of these are just thought patterns that we, for some reason, have chosen to believe. The practices of yoga, specifically the eight limbs of Ashtanga Yoga, are the remedies that we use to remove the defect of these created perceptions that bind us to a false sense of self, a sense of self that does not bring satisfaction or happiness, or fulfill our inner purpose as individual human beings. Yoga removes the defect of a mind that is attached to its own rightness.

So, to sum up our exploration of the word *yoga*:

* *Yoga* comes from the verbal root *yuj*, which means "to yoke, or to join."

* It indicates a special type of concentration, where our minds become completely absorbed in the object we are focusing on.

* Yoga is an upaya, a remedy for alleviating identification with ideas and objects other than our inner awareness.

* Relation in yoga refers to the relationships we have with our bodies, emotions, thoughts, memories, and inner sense of self and purpose.

* The meditative practices of yoga reveal our innate goodness.

* Yoga addresses the three most important questions that we can ask ourselves in our lives: Who am I? What am I doing here? What should I do next?

{ THE EIGHT LIMBS }

ASHTANGA YOGA LITERALLY MEANS the yoga of eight limbs, or parts, enumerated by the sage Patanjali. Though dating texts is a challenge due to a lack of record keeping in the ancient Indian philosophical traditions, the most recent agreed-upon dates when Patanjali wrote his text are around 200 C.E. In the Indian wisdom traditions, counting or enumerating different practices and groups of things is common; the eight limbs of yoga, the four Vedas, the 108 Upanishads, the twenty-four categories of experience, and so forth. Enumerating things helps to keep our mind organized, so we have some guidance or focus when thinking about abstract ideas. The eight limbs have traditional meanings associated with them, but in this book I take a slightly lighter and more contemporary approach. The literal translation of Sanskrit words can be cumbersome, and literalism does not always help us progress or transform, or even to understand what a word is supposed to indicate. Much of the Sanskrit canon is written in allegory form. Translations cannot be done by dictionary alone, and this is where a lot of confusion about the meanings of the Vedas, the Upanishads, and Hinduism in general gets created.

Ideally, what we are looking for in a spiritual practice is transformation, not rigidity. In this book I discuss the eight limbs of yoga in relationship to the conscious choice we make to engage with our growth, honesty, discipline, and transformation. When looked at through this lens, the eight limbs become a way for us to check in with all of our levels of engagement, from the level of the world and the people around us to the way we relate to our inner being. After all, experiencing the world is one of the primary ways that we know we are alive; we live in a world that is interconnected, interdependent, and vibrantly diverse. If we live only in our own thoughts, we become cut off from experience, where we actually live.

The classical eight limbs look something like this:

1. *Yama*—ethical codes of non-violence, truth, non-stealing, sexual restraint, and non-covetousness

2. *Niyama*—personal observances of cleanliness, contentment, austerity, repetition of mantra, and surrender to God

3. *Asana*—the practice of postures

4. *Pranayama*—the practice of breath control

5. *Pratyahara*—withdrawal of the contact of the senses with the objects of the world

6. *Dharana*—sustained concentration

7. *Dhyana*—uninterrupted meditation

8. *Samadhi*—the experience of a non-difference between the seer and the seen

While these are accurate translations, and while it is important to be aware of the use of these technical terms, it's also quite important for us to understand, as the "end users" of the practice, how these limbs can act as guides. How do I take responsibility of myself as I engage in a spiritual practice? How do I apply the limbs so that I can transform, so that I can soften my rough edges? With that idea in mind, the eight limbs can be viewed through a lens of consciously choosing alternative behaviours, and look something more like this:

1. *Yama:* I consciously choose to make my interpersonal interactions thoughtful, loving, and respectful.

2. *Niyama:* I consciously choose to dedicate myself to my spiritual practices and disciplines.

3. *Asana:* I consciously choose to take care of my body and mind through practicing postures.

4. *Pranayama:* I consciously choose to regulate and balance my breath and nervous system through breathing practices.

5. *Pratyahara:* I consciously choose to pay attention to the awareness that lies below and is the power behind my sense organs.

6. *Dharana:* I consciously choose to direct my focus and attention, and to refocus myself when necessary.

7. *Dhyana:* I consciously choose to move my mind toward absorption in my objects of focus and attention.

8. *Samadhi:* I consciously choose to shift my perception toward an experience of unity consciousness.

Regarding the first limb, ethical codes can sometimes cause mental rigidity, and have a variety of interpretations that are not always clear. However, it is important to have boundaries, and the yamas specifically help us set healthy boundaries, but we also have to make sure we are not imposing something on our psyche that we are not capable of living up to—and which then causes us to feel worse. I think that, in our day and age, it is better to take personal responsibility for adhering to these limbs in ways that are sustainable and suit our lifestyles. We can be creative with them; the only thing we have to check, once in a while, is whether or not we are being authentic with them. If we are not, usually someone will point that out to us!

The first five limbs describe observances and physical practices, and the last three describe inner experiences of deep levels of concentration and absorption. The eight limbs as a whole are defined as practices that remove the impurities that cloud our field of consciousness and lead to deep levels of discrimination, which culminate in spiritual liberation from the bondage of our conditioned minds. The goal of yoga as presented in the *Yoga Sutras* is the discrimination of the seer from the seen, the distinction of the experiencer from that which is experienced, of subject from object, so that our awareness rests within its own self, and does not get lost by identifying with the changing nature of the world. This is what is called freedom in the yogic tradition.

Historically, India has been a land of oral tradition, and even today it still is in many ways. The teachings of yoga are varied, are sometimes contradictory in detail, and differ depending on location. North Indian yoga practices, for example, are quite different from South Indian practices. Patanjali gathered the teachings that were being used in his time and earlier, and put all of the teachings together in a systematic manner.

In each of the following chapters, I will discuss various ideas related to these limbs, including scientific discoveries, psychological insights, physiologic structures, and spiritual references from texts. The first chapters focus on foundational information about the first four limbs of Ashtanga Yoga, and the second half of the book dives into science. My hope is that by the time you reach the end of this book, you will be left with a well-rounded understanding of the inner mechanisms of yoga, one that uses tradition as its base and contemporary language and scientific findings to sense its expanding relevance in today's world. Yoga is a contemplative practice—it arose from the mystical traditions of India in order to give human beings a framework for experiencing who we are and what we are doing here on this small little planet floating in space. However, the place where contemplation occurs is in our bodies, and so that is where we will begin.

three

THE PRACTICE OF POSTURES

PRACTICALLY SPEAKING, WHEN IT COMES to trying to figure out who we are and what our purpose is, our body is an obvious and easy place to start. Easy because we can see our body, we can feel it, and we can put it into different shapes. Sometimes starting with our body can be as simple as sitting down in a cross-legged position and doing nothing but breathing and listening. As children, we played with our bodies very freely, moving about, putting ourselves into different positions, standing on our shoulders or doing cartwheels—I remember standing on my head against a wall when I was five, just to turn upside down. My best friend in elementary school used to do backbends, out of the blue, that were advanced yoga poses—somehow they came naturally to him. Kids love to do backbends and spin around in circles. We express ourselves through our bodies and through movement, and movement wires our nervous system (neurons) and our brain. We are born with pretty much all of the neurons that our brains need to function, though it is true that we also grow some new neurons later in life. Our neurons begin to wire themselves to each other as we grow and develop. Each new movement that an infant learns to do creates a pathway in the brain, which also reflects how information is processed in the brain.[1] Our first movements include learning to lift our head and neck, roll over, crawl, and eventually walk. Many of these movements occur in yoga practice as well, and can help reinforce through our body how the brain processes information, which it does in the following directions:

* Back to front across the motor cortex, translating thought into action

* Up and down, from the bottom to the top of the brain, for emotional processing

* Side to side, through the corpus callosum, for comprehension

In yoga practices we find up and down and forward and backward movements in the sun salutations; in standing postures there are side-to-side movements; in seated poses and back-bending poses you also go up and down—when we decide that we need to look at our lives through a new lens, moving our bodies into new shapes will help to change our perspective on ourselves and life because we are directly using our bodies to influence the way we process incoming information; our worldview can easily be altered by putting ourselves into postures. Through neuroplasticity—or the ability of our nerves to wire together in a myriad of ways as we learn new things—we create pathways that lead us to have insights about our body, our emotions, our thought patterns, our relationships, and, most important, our sense of self. It's more than likely that this is what the yogis did when they began experimenting with postures, and that they did it with innocence, curiosity, and sincerity.

Concentration does not mean screwing the mind into a fixed state of focus, and the practice of postures does not mean forcing the body into complicated poses: both are about achieving calmness and filling the mind with a natural state of goodness, uncovering a natural, underlying trait that has been covered up by too much thinking. Postures, when done in a calm manner, help us toward this goal, because the mind and the body are not two separate things but one continuous process. The body, in yoga, is a manifestation of the mind. Our thoughts, feelings, and emotions are felt, held, and expressed through our bodies. We can see on someone's face if they are happy or sad; we can see in someone's posture whether they are feeling dejected or confident. Feelings and mental characteristics that we think of as existing in the mind exist equally in the body, just as two thousand years ago, mental states were assigned to organs and other tissues. Hippocrates is credited with developing the theory of the four humors, which, though rejected by modern medicine, is still sometimes used to describe psychological states (melancholic, phlegmatic, choleric, and sanguine). Both Ayurveda, the Indian science of medicine, and traditional Chinese medicine have similar theories. Organs, emotions, thought patterns, and temperament are interconnected. By working with our bodies, we work with the mind, heart, and emotions at the same time.

The practice of yoga postures, called *asanas* in Sanskrit, is found in yoga texts from as long as two thousand years ago. The word *asana* is made up of two parts: *as*, "to sit," and *ana*, "breath." To do an asana is to literally sit with your breath, or to sit in a special way and breathe. When you sit with your breath, you sit in awareness. While the universe is a mystery, our bodies are equally mysterious, so it's natural for our body to be the first and most obvious place to begin an inward investigation into the great mystery of who we really are and where awareness exists within us. A famous verse from the fourteenth-century yoga text called the *Hatha Pradipika* is one of the earliest sources that says postures are the first and perhaps best place for new students to begin their inward journey:

Hathasya prathamam gatvad asanam purvam uchyate
kuryat tadasanam sthairyam arogyam
changalaghavam. 1.20

||||||

Among the practices of Hatha Yoga, asana is said to be
the first practice, as it gives steadiness, freedom
from disease, and lightness of limb.

Steadiness, health, and lightness of limb: these all sound
like very desirable outcomes of a yoga practice, and show that a
lot of how we think of yoga these days was as true in the
fourteenth century as it is now. Most people associate yoga
with:

* A fit and flexible body, which is *lightness of limb*

* Relaxation, analogous to *steadiness*

* Reduction of stress, which is essentially *freedom
  from disease*

These three outcomes are correlated to the three aspects of
our body–nervous system–mind complex that we are working
with in yoga: lightness of limbs occurs in our body; freedom
from disease occurs when our nervous and immune systems
are resilient and in balance; and steadiness is what occurs in our
mind. The practice of postures is a holistic practice that affects
many layers of our being, not just our physical capabilities, and
yoga asanas are clearly not just for our bodies.

Each of the benefits listed above has outer and inner conditions, or indicators. The outer indicators of steadiness, for
example, are bodily strength, flexibility, and good proprioceptive sense, but the inner conditions for this sense of steadiness
occur when the nervous system is balanced and functioning

well. Freedom from disease means that our body can remain in a state of health because our immune system is resilient, and our physiological systems—cardiovascular, respiratory, digestive, endocrine, and so forth—are functioning properly. A strong immune system is also linked to a mind that is positive, calm, accepting, filled with gratitude, and therefore able to manage stress.[2] The lightness of limb that is spoken about is a quality of both the body and the mind, for as the body becomes lighter, or more easeful with practice, the mind becomes increasingly light as well. This verse is also an indicator that there is no body-mind distinction, or that body-mind is a continuous whole, because practicing asanas confers benefits on body, nervous system, and mind all at the same time. If you can keep these three benefits of asana practice in mind, they can help you remain focused on the inner purpose of asanas while you are doing them. It is easy to slip into doing postures just for the sake of doing postures, because the physical aspect of our being is the main thing we see and feel and where there is so much sensation. However, as Pattabhi Jois used to say, yoga is an internal exercise: what happens underneath the surface of the posture is where the lasting changes to our mind and emotions occur.

In regard to the nervous system, many of the diseases that ravage civilized society today are caused by stress and are largely preventable: heart disease, high blood pressure, diabetes, certain cancers, irritable bowel syndrome, and depression. We will discuss this at greater length in chapter 11, on the nervous system, which details the role that stress plays in mediating environmental loads (which is what stress essentially is).

THE THREE GUNAS

To understand why asanas create lightness in the body, we have to discuss some Sanskrit terms, starting with the three *gunas*. *Guna* literally means "a rope." All the things that we experience in the world, or in nature—that we see, hear, taste, touch, smell, and sense—are made up of the three gunas, as is everything in the world that can be named, or that has a form, whether visible, like a body, or invisible, like an electron. As a starting point, they can be defined as follows:

1. *Sattva:* that which is harmonious, bright, clear, buoyant, and reflective

2. *Rajas:* that which clouds perception, is active and passionate; the spark of creativity

3. *Tamas:* all that is heavy, inert, or dark; the force of gravity and giving mass to objects[3]

The gunas are often referred to as the three qualities of the universe, and are also called *prakrti*, or nature. In yoga, nature is defined in two different ways:

1. As being in equilibrium

2. As moving out from equilibrium

When nature is in equilibrium, the only thing that exists is potential, which is the potential for the universe to come into being, the potential for existence to take shape as form from the formless potential. This is called "infinite potential." When nature moves out of equilibrium, it creates the manifestation of the universe and all that we see in it, called "infinite creativity" or "infinite manifestation," and begins to create different patterns and rhythms that are essentially the operating functions of both the universe and our world. It is the gunas that are at rest in potential, and the gunas that move when they manifest as creation. They are the ingredients of the elements that make up our world: earth, water, fire, air, and space. All material creations come from a combination of these elements, including things that are invisible to the naked eye, such as atoms and photons. We'll talk about some of these patterns in later chapters. The myriad of things we observe in the universe, the infinite manifestation, are a complex, incomprehensible intertwining of the gunas. As infinite manifestation, the gunas combine in ways that produce not only this world but also all of the other worlds and galaxies that are unknown to us. As infinite potential they are at rest, containing infinite possibilities that begin to manifest when gently woken by the ripple of awareness.

We can identify the gunas in any object we can see in the world. In a traditional example, we can look at a candle: the wax is tamas (inert), the flame is rajas (it burns), and the light from the flame is sattva (it illuminates). These three are not mutually exclusive. In order for any object to fulfill its purpose, the ratio of the gunas must be balanced. If there is too much wax, and no wick showing through, you can't have a flame. If there is a long wick and too much flame, the wax will melt away quickly, and you won't get long-lasting light, plus there will be lots of smoke (rajas). If the wax is not created proportionately, it will not melt, which will prevent the flame from igniting, which would be a problem of tamas not fulfilling its supportive function.

The gunas also make up our body and mind. In the body the gunas can be seen as:

* *Sattva:* associated with the mind (reflective)
* *Rajas:* associated with the nervous system and digestion (electrical impulses and digestive fire)
* *Tamas:* associated with the muscles and bones (steadiness)

The yogi Brahmananda, in his commentary on *Hatha Pradipika*, says that the practice of postures is specifically to reduce the overactivity of rajas, or excitement, that occurs in the nervous system, and tamas, or heaviness of the body, that occurs especially in the lower belly and down into the legs.[4] Rajas shows itself in many of the compulsive problems we have in the world today, primarily our addiction to activity, whether being always plugged in, always needing to be busy, or always needing to have our attention swept up in something, whether the latest news cycle, political dramas, or other people's problems. The impulse to busyness, the impulse to sensory indulgence, all of these things are rajas being overfed. Slowing down, practicing a few postures, taking a few deep breaths, practicing meditation and relaxation—all stop feeding the fire of rajas. The experience of high stress, for example, is also a condition of rajas. The physical sensations associated with stress, such as higher body temperature, sweaty palms, increased heart rate and blood pressure, are all associated with rajas, because activity and speed create heat. As the physicist Lothar Schäfer said in his book *Infinite Potential*, "In physics heat is motion. The atoms in things are constantly in motion." When we boil a pot of water, it is the speeding up of the molecules that makes it hot. We do not sip that water and remark, "Ouch! Too fast!" We say that it is too hot. "The experience of hot and cold," Schäfer continues, "isn't of fast and slow," it is as sensory stimulus.[5] The senses translate a mechanism of nature—in this case speed sensed as heat—into an experience that gives us a false sense of reality; it is the senses that mistake sensation for reality, and the gunas are the properties of nature that can create our misperceptions. However, as we increase the power of sattva, or reflection, we begin to see through our illusory, sense-driven perception of the world. Specifically, we begin to see that there is something deeper than just our perceptions: our inner awareness.

Tamas is, among other things, the quality that veils consciousness. It can be the heaviness that comes along with mental attitudes of complacency or laziness, and is also associated with obstruction, disorder, and decay. Hindu priest and scholar Rami Sivan has said that tamas is inertia as defined in physics as a perpetually moving force, and is as well the force of gravity and that which gives mass to objects.[6] Tamas manifests in daily life, for example, as in the couch potato lifestyle or sitting at an office desk eight hours a day, which creates heaviness in the lower belly and weakness in the legs. This heaviness can lead to poor digestion and elimination and a foggy mind. Sitting, in particular, has also been shown to lead to a higher risk of having a heart attack.[7] Rajas and tamas are not the "bad gunas." They are simply properties of nature that can either conceal or excite, and we need both of those things at different times and in different circumstances. It's only when they get out of balance that they become a problem.

The reduction of tamas is also related to the strengthening of the digestive fire, and a strong digestion is linked to stronger immunity and health.[8] Asanas, bandhas (pelvic floor muscle contractions done with breathing), and pranayama all can help strengthen the digestive fire (along with a clean diet—though when your digestive health is strong, you can pretty much eat anything you like). So Brahmananda states specifically that asanas give the three effects that they do (steadiness, health, and lightness) *because* they reduce the rajas and tamas that are out of balance: the "fickleness of the mind" of rajas, and the heaviness of the body from tamas.[9] This is the primary function of postures.

One other way of looking at this is that rajas and tamas are not reduced but transformed.

* When an overactive rajas is purified, it turns into creativity and vitality.

* When an overindulgent tamas is purified, it turns into steadiness and reliability.

So rajas and tamas are qualities of nature that have a job to do, and that is to keep things moving and to keep things steady.

We can have an excess of sattva, too, which can manifest as an attachment to our intellect, and that can most definitely be problematic, leading to self-righteousness, intellectual arrogance, and an inability to listen or dialogue with people of opposing views (which occurs when tamas tinges sattva).

When the gunas first begin to vibrate, or move, from their equilibrium, they begin to knot themselves together and create the blueprint for all manifestation. As these knots grow tighter, they become fixed, similar to how gasses and gravity joined together to create the stars and planets of the two-thousand-billion-plus galaxies in our universe. They become seemingly solid. Within us, our awareness, our narrative, and our identities take the form they do because the gunas have become knotted down in a particular way. These knots are called *granthis*, and are said to be located in our nervous system and in our subtle, energetic body (we'll cover these in chapter 11). However, we have some leeway. If we have a rough edge—say, we are stubborn (tamas), or easily irritated (rajas), or intellectually arrogant (sattva)—we can soften that edge through practice and self-reflection. The rough spots in our personalities are not necessarily fixed; they can be cleaned up, smoothed out, and untied just enough for us to see that we do have the potential to change and grow at will. As you'll see in the next section, will, or conation, is ruled by rajas. We use the energy of rajas to help transform ourselves.

What are the three gunas as the constituents of our mind? Swami Hariharananda says in his commentary on the *Yoga Sutras* that

- Sattva is cognition, or knowing.
- Rajas is conation, or effort.
- Tamas is retention, or memory.[10]

The balance of the gunas determines the effect or influence that they have. Balanced gunas within our mind—the ability to know, to think, and to remember—are not only vital but are also the functions we associate with having a balanced, healthy, functional mind. And yoga, after all, is primarily for the mind.

Sattva is our ability to understand incoming information, as well as that which allows us to understand things about ourselves—our emotions, feelings, thoughts—and to be reflective about our purpose in life and the world around us. Rajas, or conation, is the effort that we use to understand things—whether about a difficult experience in our lives or the injustices of the world. It is the effort we apply to understand philosophical subjects, the energy behind contemplation, and the effort that is applied outwardly as action. Tamas as retention, or memory, is the ability to hold on to experiences. Some of our experiences get held for long periods of time as our long-term memory; some are held for only the brief amount of time that the mind, or perhaps the intellect, deems necessary.

Asanas use rajas to create vitality and use tamas to create stability. Our bodies have something called muscle memory, so via repetition, postures become easier over time until even challenging poses become normal for us. Muscle memory and experiential memories are both qualities of tamas, and we can use tamas to maximize or reinforce positive ways of both moving and thinking. One of my favorite practices that supports this positive aspect of tamas as memory retention is something that Dr. Rick Hanson speaks about at length in his book *Hardwiring Happiness*. The brain, he says, has evolved to learn more quickly from negative experiences for the purpose of survival. This is called a negativity bias.[11]

Our survival, over the ages, has depended on our ability to discern when our life is being threatened. Ignoring a potential threat could lead to us becoming food for something that we would rather have for our dinner—except, of course, if you are a vegetarian (though you still don't want to be eaten). Because of our innate impulse to live, and to survive at all costs, our brains became wired to install a negative experience into our long-term memory in a very short time period. A positive experience, as pleasant as one may be, does not help us to survive, so it takes longer for one to become embedded in our long-term memory. Although there are many different types of memory, and a variety of ways that memories are stored and retrieved depending on situations and need, I am speaking of memory here in its most general sense of how an experience stays within us, impacts our future actions, and helps shape our sense of self. Dr. Hanson teaches a practice called "Taking In the Good," where you consciously hold on to a positive experience, soaking it into your memory, and into your body, for twenty or thirty seconds—it can be either an experience you've just had or an accomplishment like staying comfortably in a yoga pose, not losing your temper, or having a kind thought about someone—and by purposefully storing it into your long-term memory you begin to change the underlying character trait of the mind, which has evolved to learn from negativity. We tilt the mental scales in our favor. I love the idea that we can train the background of our subconscious mind to hum with positivity. Asanas are also a way of practicing Taking In the Good, but for the body.

HOW TO PRACTICE POSTURES

As mentioned earlier, the practice of asanas in India dates back at least two thousand years. Like many contemplative practices, the details of the practices change over the years, but the basic principles behind them stay intact. While new postures may have come into vogue over the past thousand years or so, the idea that yoga is primarily a practice for the control of the mind has remained constant. Though asanas may not have always looked the same as they do now, we still call them yoga postures because what they essentially do is act as a container for the inner experience of self to happen. Many of the postures that are spoken of in the early texts are still practiced in the same way today as they were hundreds of years ago, but there are also newer postures that have been added to the corpus through the experimentation and adaptations made by teachers over the years. Water that flows in a river is not continuously the same water, but we still call it the Ganga, Hudson, or Nile. A river is a container for a particular stream of water; yoga is a container for the stream of *self-knowing*. So although there are yoga postures that are perhaps done differently today than one thousand years ago, it does not mean that they are not authentically yogic, if they indeed remain mechanisms for self-reflection. The sages said that the stream of self-knowledge flows from pure, absolute being and consciousness, which is infinite, and not bound by time, space, location, name, or form. Asanas are gateways to that level of being.

TWO PARTS OF POSTURE

In the late 1920s, Pattabhi Jois learned a unique method of practicing asanas from Krishnamacharya that made use of two complementary principles: *vinyasa* and *asana sthithi*. Both principles are equally important. In the early Vedas, *yoga* was used to describe an oxen yoked to a cart; in astrology, *yoga* is the word used to describe the conjunction of planets or stars; in the Upanishads, *yoga* was used to indicate control of the breath and sense organs. In Patanjali's vision of yoga, as we saw in chapter 1, the word is used to describe a special state of concentration. Like the word *yoga*, the word *vinyasa* has been used at different times in different ways. It has been used in regard to ritual in the *Mimamsa Sutras*, as a special way of sitting in the *Mahabharata*, as an entire approach to practicing yoga by Krishnamacharya, and as a breathing-movement system by Pattabhi

Jois. In this chapter I'll discuss vinyasa primarily in light of how Pattabhi Jois viewed it. Much of the yoga that is currently being practiced in the West is based on, or draws from, the teachings of Krishnamacharya and Pattabhi Jois's teachings on vinyasa.

One of the verses about asanas and vinyasa that Pattabhi Jois used to quote with great regularity was attributed to Vamana Rishi from a text called the *Yoga Korunta*. It says: "*Vinā vinyāsa yogena āsana adi na kārayet*," which roughly translates as "When doing yoga, do not do the many types of asanas without the use of vinyasa." Vamana says clearly that if you practice yoga—which, remember, is for concentration of mind—you'll be doing asanas as a part of that practice, and that you should make sure that you do them with vinyasa.

Korunta means "group," and Pattabhi Jois said that the *Yoga Korunta* contained different groupings of asanas and pranayamas. According to Pattabhi Jois, Krishnamacharya learned this text orally from Rama Mohan Brahmachari, who later told him that he could find a copy of it at the University of Calcutta library. He apparently did so, but the book was badly damaged, as many texts written on palm leaves are in India, ravaged by time, insects, and neglect. Although Pattabhi Jois never saw this text, he did indeed learn it from Krishnamacharya, though only some of the verses; the asanas and the technique of vinyasa he learned both came from the *Yoga Korunta*.[12] Perhaps one day an intrepid yogi or researcher will track down a copy of it, but until then, the *Yoga Korunta* is one of the many lost texts of India.

One of Pattabhi Jois's contributions to Krishnamacharya's

teachings was to organize the "mountain of asanas" that Krish-namacharya taught him into sequences of poses that he categorized as Yoga Therapy, Nerve Purification, and Strength Portion (*Yoga Cikitsa*, *Nadi Shodana*, and *Sthira Bhaga*). In Krishnamacharya's first book, *Yoga Makaranda*, not all of the asanas are presented in an organized way. For example, the first asana is a forward bend holding the toes, and the second asana is a deep backbend, where one arches back to grab the calves! This could be attributed to the instruction in the book that yoga should be learned from a teacher, so the asanas presented could be primarily for demonstration purposes.[13]

VINYASA

In the verse of Vamana's quoted earlier, the word *adi* occurs. *Adi* means "et cetera," or "the many different types," so Vamana must have known, at whatever point he wrote this text, that there were many different types of asanas that one could practice—but that all of them should be done using the same technique. This is our first clue to the meaning of vinyasa: it is a technique. But a technique of what? As with most Sanskrit words, it's never just one thing. Vinyasa can be:

* The linking of breath and movement (this also describes the way can we move into and come out of a posture)

* An entire methodology, or approach, to practicing asanas

✳ A ritual, or methodological approach, to a
devotional practice

In Ashtanga Yoga, the first description of linking breath
and movement is the one Pattabhi Jois used, but the other
two are equally true. For example, the entire approach to practicing asanas in a way that they are linked together was called
vinyasa krama by Krishnamacharya (*krama* means "a sequence"
or "a path"). Depending on where you look, you can find slightly
different uses of the word *vinyasa*. In Pattabhi Jois's practice,
we have to adopt it in the sense that he used it, the linking of
breath and movement that is used for each position leading
into, and out of, every asana. It is *not* the flowing of one posture into another—it is *the movement of breath* that occurs
within each discrete asana or transition position.

I say this because the word *vinyasa* has of late become synonymous with "flow," as in "Vinyasa Flow Yoga," and the flow
that is described in this type of yoga is a flowing from one pose
to another not always coordinated with breathing. In Pattabhi
Jois's vinyasa, breath occurs as the body moves into, or out of, a
single position, but is not concerned with flowing into the next.
This is a subtle point, but an important one, because when you
are concerned with flowing into the next pose, your mind is
looking to the future. When you are concerned with the breath
within one posture, or one movement, your mind is in the
present moment, which is where yoga wants us to be. Vinyasa
as a breathing-movement practice is one breath and one movement at a time. To do one thing at a time is the hallmark of
attention.

Vinyasa is a breathing and moving system that frames the entirety of the asana practice, from start to finish. However, it is only part of the picture. Static asanas, *asana sthithi*, are the other part and are equally important. Contained within the breathing-movement system are a whole host of details, subtleties, and awarenesses generating techniques that become more pronounced over the years. We don't have to worry too much about learning the details early on because they reveal themselves to you as you practice and as your awareness becomes more subtle. The things you pay attention to will, over time, teach you what you need to know, so that much of what we learn will be personal and unique to our needs, though some of the discoveries will overlap with other people's experiences as well. This common overlap of experience is why yoga is called a universal practice; no matter who does the practice, many of the effects, benefits, and realizations are the same. The guidelines are what we need to learn from a teacher; we fill in the experience through practice. Etymologically, *vinyasa* is made up of three syllables, *vi*, *ni*, and *as*. *Vi* is for *vishesha*, which means "special"; *ni* is for *niyama*, which means "rules"; and *as* is the verbal root that means "to sit." *Vinyasa* in its most etymological form means "the rules for sitting in a special way."

VINYASA AS RITUAL

Ritual makes up much of our life, whether mundane or sacred. The way we brush our teeth, prepare for work or school, greet each other, start or finish our day: our lives are made up of small rituals. Religious and spiritual traditions observe daily, monthly, and yearly rituals. Our rituals include our relations to ourselves, the people around us, the planet, and our sense of the sacred. (While a habit and a ritual are closely related, a ritual is something that we do with awareness, or is something that helps to generate awareness, while a habit can be done mechanically.) Yoga, in its most basic sense, is a ritual that we perform to help us remain established in awareness. The daily practice and mental approach that we take to doing it is a certain type of ritual that eventually becomes part of the rhythm of our daily life.

Stepping onto our yoga mat each day can be viewed through the lens of ritual. It is a sacred time for us to commune with our body, our breath, and all of the things that occur in our mind—memory, emotions, thoughts, feelings, goals, and ambitions. It is a time to reach toward our inner sense of self, and to try to maintain inner quiet. Ritual sets the atmosphere and scenario for inner communion; the practice is the technique that we ride to harness our awareness inwardly. This can be done with posture, with breath, with a mantra, with a religious ritual, or simply with our own sense of awareness. There is an important ingredient to spiritual practice called *bhavana*, which means "a mood or feeling that is associated with an action." The feeling of clarity, exhilaration, or peace that goes along with your practice is something that you can consciously reinforce, so that every time you inhale and raise your arms over your head you associate it with a feeling of peace, inward focus, joy, or whatever it is that you feel. Bhavana is the conscious, repeated reinforcement of an action with a feeling, mood, or emotion. The same holds true for repeating mantras or performing other rituals. The bhavana associated with the practice is the juice or nectar that fuels it, that fills you with a presence of awareness within your mind and heart. If there is no bhavana, then the practice can become mechanical, and can become a chore to do, or we can lose interest in practicing. It is the mood or emotion that we consciously bring into our mind and heart and that we link with a physical action that makes yoga a spiritual practice. This is much more important than the ability to do difficult postures or know a lot of anatomical details about how to perform a posture.

When we plan out the ritual of our day and lives, the sequences of events that we hope will take us from point A to point B, we can think about the idea of sequencing that Krishnamacharya spoke about later in his life. He sequenced postures to create a particular effect, but also to make sure that they were varied, they complemented one another, and they didn't all move in one direction. Each day we should move both backward and forward, twist, turn upside down, and create some strength. Variation in life is good; a little chaos is healthy. Just like we want variation in our heartbeat, and differentiation of our cells, we want life to shift a bit here and there, too.

HEALTH BENEFITS OF VINYASA

Here's one theory behind vinyasa as a healing modality. Pattabhi Jois used to say that yoga was "internal cleansing" of the body and mind, because both are like sponges; we soak up the food, drink, and thinking from the environments we live in, just like a sponge soaks up liquid or dirt from a table or floor. I sometimes compare the cleansing process of yoga with wringing out a dirty sponge. In order to clean the sponge, you hold it under the tap, and twist it first in one direction, and then in the other, to wring out all the dirt. Finally, as the water from the tap rushes through the wrung-out sponge, all that passes through is clean water and no more dirt. In asana practice, we do a similar thing. First we bend in one direction, then in the opposite: pose and counterpose. First we twist right; then we twist left. We are methodically wringing out stress and stiffness from our bodies, and clearing the mind in the process.

Each time we bend one of our joints in a posture, such as in tree pose, or in a forward bend with one of our legs bent, the blood circulation at the bent joint becomes temporarily impeded. Then, upon straightening the limbs, the blood circulation flows fully again, similar to, but of course not exactly like, clean water through a sponge. In this way, through the entire course of asanas, each of the joints of the body is bent and straightened, slowing and then releasing the blood flow. The coordination of movement and breath warms the body, opening the arteries and capillaries, allowing for the blood to flow more freely, and the capillaries expand, allowing for a greater surface area to be created for gas exchange.[14] The feeling is invigorating, and it is also good for our health. The increased blood flow is also beneficial for healing a variety of pains or injuries that can occur in the body.

Vinyasas have three primary benefits:

1. They create heat in the body, which is cleansing.

2. They create strength of the breath, which helps us to concentrate.

3. They create the *appearance* of a flow, which can be absorbing for the mind.

Vinyasa describes the action of synchronizing breath and movement. Breath and movement are a natural pattern of the body, nervous system, and respiratory system, and vinyasa maximizes the effects of this pattern. Our bodies are a collection of physiological processes that all follow patterns; likewise, our bodies and breath are in a constant pattern with each other.

For example, when we inhale, our chests expand and the belly will protrude a little. When we exhale, our chest relaxes down and the belly might come back in a little. You can observe this easily in a sleeping baby. With vinyasa, we are taking this natural breath-body movement pattern, this up-and-down pattern, and applying conscious awareness to it, whereby every inhale will be a conscious upward movement, and every exhale will be a conscious downward movement. The breaths we take, either in or out, are done in conjunction with moving into discrete asanas such as *chaturanga dandasana* (the push-up position) or *urdhva mukha svanasana* (upward-facing dog), or in transitions, such as when we lift our heads and chests up before and after a forward bend.

I sometimes like to think about vinyasas, or linking movements, as the individual frames of a movie. Each vinyasa is like one frame. When we watch a movie, it takes on the appearance of a flowing sequence of events, sort of like real life; but when we examine the film, we see each distinct frame. When we apply awareness to each vinyasa, we can, for the time we are practicing, live in a frame-by-frame, or moment-to-moment, existence. In this mental space it becomes easier to examine our lives, goals, ambitions, faults, and tendencies. The vinyasas, in this way, can teach us to live in the present, one moment and one breath at a time.

However, there is another level of vinyasa, and this occurs when the practical details become second nature to us, and the frames become internalized and natural. Our awareness can then expand beyond just that one frame and spread out to the whole process; we enter into a state that feels like we are in a flow, both body and mind, whereas earlier the mind was focused on just one frame at a time. Both are important, and there is going to be a coming and going of how we move in and out of each state. You have to practice scales before you can play a sonata, and you have to repeat many mantras before you can become absorbed in pure sound. Likewise, we have to practice many asanas, and many vinyasas, in order to enter into the flow of awareness.

In chapter 3 of the *Yoga Sutras*, Patanjali describes concentration (dharana) as the mind's ability to become completely fixed in one place; however, the thought process of dharana is not a steady flow of attention, but is intermittent. The analogy given is like successive drops of water, perhaps from a dripping tap, that follow one after the other. Concentration becomes meditation (dhyana) when the flow becomes continuous and uninterrupted, like honey being poured from a pot. Vinyasa is the initial training of the mind to return again and again to a point of attention. The difference between vinyasas and the meditative concentration of dharana is that in concentration we choose only one object to return to again and again, while in vinyasa we go from one point to the next, such as one asana to another, an upward movement to downward, etc.—nevertheless, both are ways to train the mind. In his commentary on the *Yoga Sutras*, Hariharananda says that bhavana leads to dharana.

When we watch someone practicing any physical activity with mastery, from the outside it looks as if they are in a flow— Ashtanga Yoga, indeed, looks like it flows. But that flow comes from the mastery of each step and, in Ashtanga Yoga, from each vinyasa. From far away, it looks like a flow; from close up it looks like discrete asanas strung together with breath, one movement and one breath at a time. When an athlete enters into the flow state, he or she is indeed in something that feels like a flow. But how did the athlete get there? By learning to be fully present in each step, each note, each swing, each vinyasa. We enter into the flow state after thousands and thousands of single, practiced moments.

The basic principles we should keep in mind in regard to our terminology are these:

* All the vinyasas lead into and out of asanas.

* When we are moving and breathing, it is called vinyasa.

* When we stay in a pose for a few breaths or longer, it is called sthithi, which means "to stand, place, or remain."

In this chapter, we have looked at the purpose of asanas, and the meaning of vinyasa as the active technique of transitions that move into and out of the static positions, the asana sthithi. *Vinyasa* does not mean flow; it is an activity of breath within movement. This is a really important, key understanding. Vinyasa brings us from point A to point B, and then back again. It's a feedback loop. A flowing river brings us in only one direction.

four

THE SEAT
OF AWARENESS

NOWADAYS, ASANAS, OR POSTURES, are one of the most visible parts of yoga, and when people say they do or teach yoga, they are usually referring to asanas. The final position of any asana, whether you are flexible or not, is called the "state of the asana," or, in Sanskrit, the asana sthithi. *Asana* literally means "a seat." Traditionally, the word *asana* was used in ritual or meditation to refer both to the position you sit in and to the grass mat you sit on for worship or meditation. Etymologically, it is made up of two syllables: *as*, a verb that means "to sit," and *ana*, which means "breath" or "respiration." An asana, therefore, is literally a position in which we sit breathing, or sit with our breath. *Sthithi* means "situation, state, position, or abode," and it is in the asana sthithi that the effects of each posture are compounded. Asana sthithi and vinyasas are complementary; all vinyasas are active and lead toward and away from asana sthithi, and asana sthithi is the grounding, static counterpart of vinyasa.

Further, asana sthithi has two components: steadiness and comfort. Steadiness is *sthira*, and comfort (or happiness) is *sukha*. In the *Yoga Sutras*, the verse that describes the practice of asana is actually describing asana sthithi, and it says, very simply:

Sthira sukham asanam. 2.46

||||||

Asana is steady and comfortable.

In every posture that we do, there should be some measure of steadiness and happiness in the pose, and these are both physical and mental components. If a pose is physically challenging but your mind is calm and at ease, then you can have steadiness in the pose. *Sukha* literally means "good (*su*) space (*kha*)." If you recall the description of the gunas in chapter 2, sattva guna, which is lightness and reflection, is equated with the trait of mind that is calm, clear, spacious, and reflective. The element of space is equated with the mind, just as earth is equated with the bones (like tamas), and fire is equated with the nervous system and digestion (like rajas). When we read that a posture should be performed in a way that is sukha, or has good space, it's a hint that we are talking about not just physical happiness but also the conditions that create happiness, which are psychological as much as they are physical, and this is a condition that includes openness, freedom, and acceptance. Just as space is a container for objects, the mind is a container for thoughts, emotions, feelings, memory, and information. The neuroscientist Dr. Daniel Siegel, one of the leading experts on the exploration of the ineffable thing we call "mind," has described the mind as the inner, subjective experience we have of life. In *Brainstorm: The Power and Purpose of the Teenage Brain*, he discusses how the mind regulates the flow of energy not only within us but also in our relating to others. In order for this to happen, the mind also has to have self-regulatory and organizational capacities to monitor the change and flow of information that is constantly coming in. Siegel sums up the content of mental experience with the acronym SIFT: sensation, information, feelings, and thoughts;[1] all of these things are flows of information that occur in the container of the mind. The mind itself is a neutral space. It is what we fill the mind with that will determine our happiness in life.

So what is steady and happy and filled with good space in an asana? Both our body and our mind. Steadiness, stillness, and joy are qualities of both the mind and the body, and it is crucial to remember that when practicing asanas. Asanas are for both improving the facilities of the body and stilling the mind, because the mind and the body are a continuum. Therefore, that which brings stillness to the body and the mind is considered to be yoga. In asana sthithi, we practice stillness. In vinyasa, we practice movement. If everything were vinyasa, we would never become quiet, so therefore the vinyasa needs to move toward stillness, toward a quiet point, where the mind can become absorbed in the present.

One morning in 1992 in Mysore, during an after-practice question-and-answer session with Pattabhi Jois, a student asked, "How do you know when your posture is correct?" He replied, "When the mind is quiet, the asana is correct." It was one of my favorite things that he ever said (and the only time I ever heard him say it), because of its simplicity and truth. Yoga uncovers the deep quiet within a mind that is dominated by thoughts throughout almost all of our waking and sleeping moments of the day. The quieting of the mind occurs in stillness. When we arrive at that place which is free from discursive thought, and where our hearts are filled with a feeling of peace, gratitude, and love, we know that the yoga is working.

Here are a few things that you can do to help move in that direction while practicing asanas:

* After coming into the position, don't fidget; you do not have to try to go deeper or further. The striving to go deeper is just another thought.

* Move your awareness through the different parts of your body while remaining steady in the posture; your body will tell you where you need to be if something needs to be adjusted.

* Keep your mind soft and relaxed; once you have achieved as much of the position as you are going to, your awareness should remain on even breathing.

Pattabhi Jois once explained that if there is too much attention paid to the posture, the quality of the breathing will be lost, and if there is too much attention paid to the breathing, the quality of the posture will be lost; therefore, we should pay an equal amount of attention to both posture and breathing.

Not all yoga postures will be comfortable the first few times (or months, or years) that we do them. But over time, our body and mind will adjust to them, and that which was difficult will become second nature. This comes from practice. The ability to stick with something that is difficult, and to do it with a calm mind, is one of the primary principles of yoga practice. Through gradual mastery we develop an ease of effort, and by keeping the mind focused in a relaxed way on the breath in difficult postures, we learn to bear difficulty. According to Patanjali, this is one of the principal benefits of practicing postures: that through them, we will be able to withstand the ups and downs of life with a calm mind and with fortitude. This is also called resiliency, the ability to bounce back from hardship or stress, without getting too thrown off balance. Drs. Steven Southwick and Dennis Charney have described this ability in their book *Resilience: The Science of Mastering Life's Greatest Challenges* as the capacity to bend without breaking, and return to our original shape or condition, and as an ability we can gain through practice (such as through asanas). Yoga is indeed a balancing act. Though *sthithi* means "stable" and "steady," and implies a type of balance, we should never forget that balance takes a tremendous amount of energy—some even say that we are never in balance, but always in the process of balancing. Our body's ability to maintain homeostasis, which includes the balance of gas exchange, glucose, temperature, and blood pressure, is a twenty-four-hour-a-day job that requires a tremendous amount of energy. Postures, breathing, and meditation practices, through a complex confluence of inputs, support and encourage the body to find balance through its inherent self-regulatory mechanisms.

THREE PLACES OF ACTION

Within the practice of postures, there are three areas that we must pay attention to: the body, the breath, and the mind. To put it very simply, postures steady the body, breathing steadies the nervous system, and gazing steadies the mind. In Sanskrit this is called *tristhana*. *Tri* means "three," *sthana* means "place." When these three basic factors are addressed, we are creating a holistically integrated body-mind complex practice. For the most part, it is not reasonable to expect we can focus on all of these equally, since the mind can really be at only one place at a time. So these three places where actions occur in yoga postures are more like reminder points, or check-in places, so that we can move our awareness back and forth between our body, which is in a posture, our breathing, which should be free and steady, and our mind, which often wanders but can be focused through different gazing points, called *drishti*.

So far we have covered the posture part of tristhana, asana, which is for the health, stability, and proper functioning of the body. Now we will cover the breath, which is for the balance of the nervous system, and dristhi for the steadiness of the mind. Breathing has other important benefits aside from just balancing the nervous system. The following list is not comprehensive, as the benefits of breathing are a huge topic, but here are just a few important points about how breathing enhances our experience of yoga:

* It is a measure of our energetic output.

* It is an indicator of endurance.

* It will tell us when we have done enough.

* It is an anchor for our awareness.

* It can calm us down or energize us.

* It creates profound mental focus.

The type of breathing that Krishnamacharya (and then of course Pattabhi Jois and Desikachar) taught students to use while practicing postures is a slightly restricted method of breathing, whereby you produce a whispering sound or hissing sound through your nose. It is the sound you would make if you were exhaling out of your mouth to fog a window, or if whispering but directing the breath through the nose instead of the mouth. The anatomical structure that creates this sound is called the glottis, which is the opening of the vocal cords at the top of the larynx, which sits above the trachea, or windpipe. During quiet breathing, the glottis is more or less open. When we speak or sing, the glottis will vibrate open and closed to help create resonance along with the voice box. When practicing voiced or vocalized breathing, the glottis is held slightly closed. Ideally the inhalation and exhalation are an equal length, and the sounds of the inhalation and the exhalation are roughly the same. Further, the breath should be *dirgha*, which means "long," and *sukshma*, which means "subtle." If the breath is too short, or we are huffing and puffing, we have lost the thread of the breath and the thread of yoga. Dirgha and sukshma indicate there is a smoothness and steadiness to the breathing within the asana practice. The idea of dirgha isn't the same as volume, so it is not that we are taking deep breaths; we are just extending the natural length of the breath.

You can also create the vocalized sound of the breath simply by consciously extending your inhalation and exhalation. The simple act of lengthening the breath closes the glottis slightly, so that you have control over its length. Pattabhi Jois called this "nose and throat breathing," because there is a conscious awareness from the nasal cavity to the trachea, and "breathing with sound." He further said that the breath should be focused between the heart and the throat, which means the lower abdominal muscles stay controlled to support the movement of the diaphragm outward, giving fuller expansion of the lungs. In trauma-informed yoga this type of breathing is also called the "ocean breath," because the sound that is made is similar to waves breaking on the shore and pulling back into the sea.

Once we have this sense of a steadiness or constancy of the breath, we can use it to measure many of the things on the list on the previous page. For example, if we are forcing a posture or struggling to get into a position we aren't ready for, the breathing will change: it might stop, it might sound strangled, or it might become very fast. This is energetic output. If we are huffing and puffing, or racing to catch our breath, then we have come to the limits of our endurance. Sometimes when we have done enough, the breath just kind of peters out, and there is no more sound being made, but we aren't gasping for air. This means that we have done enough for the day and should rest.

The steadiness of breath along with the postures also helps us to focus our awareness in the present. When the breath is steady, then the asana is steady and so is the mind. When the breath is unsteady, shaking, or is being held unconsciously, then both our effort and mind are probably not calm. Through practice it is possible to use an energetic effort in order to achieve a challenging posture while keeping your breath steady, which is an indication that your breath, body, and nervous system are getting stronger. As well, the breath can both energize and relax us. Through strong and steady breathing we can create energy, and through relaxed, extended exhalations we can promote relaxation. In chapter 11 we'll go deeper into this.

Finally, when we are fully focused on the breathing, our awareness moves inward to be fully present with each movement we are making, each breath we are taking, and the inner sensations created by yoga practice. This creates a deep level of interior awareness and focus, which is the hallmark of yoga. Vocalized breathing also creates friction, which means that it naturally warms us up. This warmth, combined with the movement of vinyasa, creates heat in the body, which is said to be purifying.

As a recap of the breathing, the breath should:

* Be of equal length on the inhalation and exhalation

* Be made with a light hissing sound in the nose and throat

* Be focused from the heart to the throat, meaning that the belly does not extend outward when you inhale

* Make roughly the same sound on the inhalation and the exhalation

Drishti is the third ingredient of tristhana, and is for steadiness of the mind. *Drishti* is a Sanskrit word that means "vision, sight, point of view," or, as a verb, "to fix the eyes on." The verbal root *dr* means "to place, or to hold," and the way that the eyes are used in yoga is such that we place our gaze on a particular point, and allow that point to hold our gaze. When we look at the thumbs or at the nose, we are not staring at that point but resting our gaze on the place that we are looking at, like placing a book on a table. When we rest our gaze on any of the focal points, the mind should come to rest, for the state of attention comes when the mind is at rest, and not when the mind is trying too hard to focus. Softening the eyes, especially the outside corners of the eyes, can have a mentally calming effect, as the nerves that innervate this area are from the parasympathetic branch of our nervous system, which rules rest, relaxation, and repair.

There are nine drishtis, or places of focus, taught in the *Yoga Korunta:*[2]

1. The tip of the nose, *nasagra*

2. Between the eyebrows, *brumadhya*

3. The navel, *nabhi chakra*

4. The hands, *hastagra*

5. The feet, *padagra*

6–7. To both sides, *parshva*

8. The thumbs, *angushta*

9. Upward looking, *urdhva*

It is a basic function of our visual apparatus that our mental attention will follow where our gaze is directed, or, to put it more simply, "the mind goes where the eyes go."[3] Try this simple experiment: When you are about to have a thought, or to recall something, or are stressed, keep your eyes very still, and observe what happens to your thinking process. You'll notice that sometimes when you begin to think, to search for a word, to formulate an idea, or to perform a math operation, your eyes will look upward toward the right or left, in an effort to access information, or the pupils will begin to dilate. Mental effort and pupil dilation have a direct correlation.[4] When we soften the eyes and keep them still, our mental activity changes along with the eyes. Softening the eyes, relaxing the eyes, and placing our palms over our eyes are all ways of reducing stress and relieving mental tension. During yoga practice, we consciously use the eyes to practice a calm, focused attention that can help support mental steadiness and hopefully prevent mental stress that naturally builds during the day.

The retina is the part of the eye that translates light into nerve signals and allows us to see everything from the light of distant stars to daylight, to perceive the various wavelengths of light as a full range of colors, to perceive both a speck of dust and a huge mountain range, and to perceive depth of field. Because the retina grows out from the brain in a direct line of brain cells, by focusing the eyes—that is, by practicing the drishtis—we are engaging a direct link to our brain. Stabilizing the vision is steadying for the mind and is said to be strengthening for the eye muscles, but perhaps most significantly, it could be one of the reasons why the yogis believe that drishtis help to control the mind, because our eyes are anatomically connected to the brain, where thought is processed. Our eyes literally grow out from our brain, and our eyes are one of the vehicles by which the brain comes into direct contact with the world around us. Therefore, visual processes have a profound impact on brain development, and drishtis use the visual processes to support coherent brain functions such as memory, perception, and the ability to practice mental stillness.[5]

During the day, when you feel tension or stress arising, you can check in with your gaze, the same way you might check in with your breath:

* Do you notice any tension in your eyes?

* Can that tension be relaxed by palming your eyes?

* Can you relax your eyes by closing them slightly and resting your gaze on a steady object?

* Does your mental focus increase when you gaze steadily in one direction?

See if the practice of drishti, and of resting the eyes so that the mind becomes restful, can be integrated into your tool kit of stress reduction or mindfulness practices. You can use them at any time of the day, whenever you need them, but they are particularly important during yoga practice because they help to keep you on point. It is not of much use to be in a yoga posture but looking all around the room, letting your attention wander. The gazing points are another tool to fix your mind in the present.

HEALTH BENEFITS OF YOGA

By now, we all know how important it is to move a little bit every day, and to exercise a few times per week. Health, in a most general sense, equals movement. When our blood is flowing properly, when our nervous system is firing clear messages, when we are digesting the food we eat, and when our limbs can move freely, we feel that we are in a state of good physical health. Our bodies are made up of somewhere in the area of thirty-seven trillion cells, which are transmitting messages to one another in order to keep our bodies—which are basically a cellular environment—functioning. There are an untold number of physiological processes happening within us at every second of the day, and we—that includes doctors, neuroscientists, biomechanical experts, and quantum physicists—have no idea what they all are, and how they all work. Our cells know a lot more than we do.

But one thing we do know is that movement somehow ties our cellular environment together. Our heart beats, our blood produces pressure, our breath flows, our bowels move, our nervous system fires, our mind jumps from here to there, and our limbs like to move. We are moving beings, so it is no wonder that daily movement will help us feel like we are being put back together. Dr. John Ratey, in his book *Spark*, writes at length about the brain-body connection, and his list of things that movement affects is profound. In an early chapter on learning, he elaborates on neurophysiologist Rudolfo Llinàs, who said in his book *I of the Vortex: From Neurons to Self*, "That which we call thinking is the evolutionary internalization of movement." Dr. Ratey says, "As our species evolved, our physical skills have developed into abstract abilities to predict, sequence, estimate, plan, rehearse, observe ourselves, judge, correct, make mistakes, shift tactics, and then remember everything that we did in order to survive."[6]

Beyond this, I would add to the list that we use these things not just to survive, but to transform. Our brains have evolved as we have evolved—simultaneously, or because of each other— into more complex, physical, moving creatures, and the practice of postures, perhaps, is helping us to evolve into more thoughtful, compassionate, reflective creatures, not just moving ones. However, it is clear from his research that movement helps to organize the functions that the brain processes, and this is why the practice of postures helps to make us feel grounded, centered, organized, and at home in our bodies.

Regular exercise not only improves our brain health and fitness but also affects many of the functions that our brain oversees:

* Stress

* Anxiety

* Hormone function

* Sleep

* Digestion

* Blood pressure

* Sexual appetite

* Body temperature

* Mood

* Learning capacities

* Memory

* Executive functions

* Expression of compassion and empathy

* Aging

Each of these facets of our existence contributes to our physical and emotional health, and they all have brain correlates. Several of them, like blood pressure, body temperature, digestion, and sexual appetite, are overseen by the brain stem; they are our survival functions. Others, such as learning, memory, and the executive functions, are processes of what we call "higher" brain functions. Executive functions are things like long-range or strategic planning, social interaction, and expressions of compassion and empathy. Several recent scientific studies have shown that the practice of postures can be helpful for supporting and improving many of the functions on this list.[7]

Over the past years I have designed yoga protocols for research led by Dr. Marshall Hagins.[8] Two studies in particular, one study on pre-hypertensive conditions in African Americans, and another on grade point average in high school students, are noteworthy because of their outcomes. The protocols were not radically different, but the audience they targeted was. Both studies had positive outcomes: in the blood pressure study, there were statistically significant decreases in systolic blood pressure; in the grade point average study, students who had forty weeks of yoga in comparison to forty weeks of gym classes saw their GPA collectively rise 2.7 points.

How is it that basically the same practices have such different effects? My suggestion is that yoga, through postures, breathing, and focused attention, effectively balances the brain functions and nervous system, and through homeostasis, the nervous system knows what needs to be balanced or corrected at any given time. When the brain is given proper support through exercise (because the brain and movement are synonymous), the functions that are being highlighted will automatically begin to strengthen. So, when blood pressure is off, yoga will help to restore the homeostatic functions that keep blood pressure at normal levels; if stress levels are high, yoga practice will down-regulate, particularly through breathing, the parts of the brain and endocrine system that are responsible for hormonal release of adrenaline and cortisol.

While it is not a cure-all for every disease, what yoga essentially does is act as an internal balancing act for our brain, nervous system, and cellular mechanisms. It restores the functions that are out of alignment toward a state of balance. The body is intelligent; it knows what it needs to do. If we behave and live in such a way that we support that intelligence, the body will restore itself to balance. If we have a disease that is not a dysfunction of our lifestyle, diet, and habits, we can still use yoga and lifestyle to help support the body's healing capacities. This can be seen in cancer or multiple sclerosis patients who use diet, yoga, meditation, sleep, and positive emotions to help them improve and manage their illnesses. In a nutshell, yoga supports self-regulation.

POSTURES IN THE PAST

Patanjali, in his treatise, does not give specific or elaborate instructions on precisely how we should do the practices that he lists. His sutras describe the names and types of practices, and what their effects are; the *Yoga Sutras* are not an instruction manual in the same way as when we buy speakers for a stereo, we are told exactly what to plug in and where, and if the speakers don't work, how to troubleshoot. Patanjali does not go into much detail because he is categorizing and collecting different practices, not telling us how to do them. For example, Patanjali says that we should practice *ahimsa* (non-harm), and tells us that the effect of non-harm is that hostility ceases in the presence of one who is established in it. But he does not tell us how exactly to practice non-harm. Does that mean we need to be vegetarians, or vegans? Does it mean we have to be non-harmful in every single situation we find ourselves in? What if we have to cause harm in an act of self-defense, or to protect someone who is being abused? Patanjali does not say. Instruction on how to practice non-harm comes from a teacher who has experienced or practiced it. It also comes from us meditating on the idea of non-harm, and then incorporating that into our own lives. We can also look to the texts, for example, the *Mahabharata*, or in the writings of Mahatma Gandhi and Dr. Martin Luther King Jr.

When it comes to asanas, Patanjali spells out in three sutras a comprehensive overview of the entirety of asana practice. The first sutra gives the definition of asana, the second sutra tells how every asana should be practiced, and the last describes the results of asanas properly performed:

1. Asanas should be done in a steady and comfortable manner. (*YS* 2.46)

2. Our effort should be calm, and our mind absorbed in infinite consciousness. (*YS* 2.47)

3. The result of practicing in this way will be reflected in our ability to withstand the ups and downs of life, or, as he put it, to transcend the pairs of opposites. (*YS* 2.48)

What Patanjali does not say are the poses we can practice. Those need to be filled in by a teacher, and can be remarkably varied, depending on the teacher and his or her experience of yoga.

On occasion it is said that because Patanjali wrote very little about asanas, they were not considered an important part of practice, or that the focus seen on postures over the past one hundred years or so is a relatively new phenomenon and could not possibly have been how yoga was practiced a thousand years ago, or even five hundred years ago.[9] This is a logical fallacy. Very little was being recorded five hundred or one thousand years ago, so how can we say that what is being done now, or what was taught by Krishnamacharya, is not what was done then? Though this claim could certainly be true for "gym" yoga and modern hybrids, it cannot be confidently stated about Krishnamacharya. Further, this kind of logic demonstrates a misunderstanding of what a sutra text actually is supposed to provide one with—which is guidelines, not explicit instruction.

And if you look closely at some of the commentaries in the yoga texts, hints are given that there is much more that is not being written down. For example, Vyaas, who was the first to elaborate on the *Yoga Sutras* in commentary form, listed eleven important asanas that could be done, but indicated by the word *adi*, which means "etc.," that there were many others. Shankaracharya, the root guru of the non-dualist philosophical tradition Advaita Vedanta, elaborated on this by saying, "By the use of the word 'adi' one should understand any other posture which has been taught by the teacher."[10] This indicates that the existence of a variety of postures was known well over a thousand years ago, and that instruction by a teacher was where the different asanas could be learned.

For those interested, there are other texts, such as the *Kapala Kurantaka Hathabhyasa Paddhati*, published by Kaivalyadham in India, an undated text that is estimated by the director of research at Kaivalyadham, Dr. M. L. Gharote, to have been written as early as the fourteenth century,[11] and dated by the Sanskrit scholar Jason Birch as a seventeenth- or eighteenth-century text.[12] It lists 112 different postures. All of the known early texts list just a handful of postures, but they also list just a handful of pranayama practices, and as Shankaracharya says in the *Yoga Taravali*, there are thousands of types of pranayamas, but he speaks only of the one called *kevala kumbhaka*, which he (and the yogis) considered to be the most important, and does not elaborate on all of the others.[13] We should not, therefore, assume that a truncated list should be equated with a limited number of techniques, but rather that it could be attributed to the nature of the medium: even if there are thousands or even hundreds of techniques, it may not be practical or time efficient to write them all down, as writing on palm leaves takes a long time and, as well, these things should be learned from a living teacher. Instead, just write down the important ones. This is why the word *adi* is added to Vyaas's list. The practice of asanas as part of a spiritual discipline is not new, their number is quite large, and types vary.

THE FOUR PATHS OF YOGA

There are four general paths of yoga:

1. *Jnana Yoga:* the yoga of inquiry, textual study, and debate

2. *Karma Yoga:* the yoga of service, and offering all actions to God

3. *Bhakti Yoga:* the yoga of devotion, through service, song, and ritual

4. *Raja Yoga:* the Royal Path, the yoga of practice (synonymous with Ashtanga Yoga)

All of these are huge topics in and of themselves, but generally speaking, each person has an inclination that draws him or her to one or to the other. Some people are more devotional in nature, while others are more analytical. The different systems exist to fulfill our individual needs or nature, called *adhikara* ("eligibility") in Sanskrit. The different approaches, however, are not mutually exclusive. Even when you are engaged in Raja Yoga you will have bhakti, or a sense of devotion, to your practice and teacher. When practice is done as an offering (karma), and with knowledge of its philosophical basis (jnana), the benefits will be all the more profound. So in reality, the types of yoga are interlinked. Each of us will choose one as our main focus, the one that is comfortable and natural and helps us to address our innermost being.

To sum up asana:

* A posture should be stable and comfortable.

* Our efforts should be done with a calm mind.

* Our attention should be focused inward, on expanding awareness to all parts of the body, not just the part that we feel is stretching, strengthening, or doing the bulk of the work.

* Asana sthithi is one of the two components of postures, with vinyasa being the other.

five
||||||

{ WHERE IS MY MIND? }

THE MIND-BODY DUALITY QUESTION has been plaguing philosophers since the dawn of philosophizing. Both the Eastern and Western wisdom traditions have long held that the mind-body is a unified field of experience; however, since the early 1600s and the age of science and reason, the West has adopted a mind-body split that gradually crept into our collective consciousness as mysticism fell by the wayside and the dogmas of religion and science began to take its place.[1] Perhaps this was cemented by Descartes, whose mechanistic and dualistic view of the world ushered in the age of reductionism, the view that the whole could be understood by taking it apart. According to Descartes, mental states were distinct and different from physical states, which has led to a mind-body split over the ages that has caused a lot of confusion and psychological distress, as people give more importance to experiences that occur in their minds but often forget that they have a body, which is also a field of experience. The British philosopher Gilbert Ryle countered this doctrine, calling Descartes's model "the ghost in the machine"[2] and claiming that the doctrine was in conflict with what we actually know about the body and the mind, that they are a continuum. While Descartes famously said, "I think, therefore I am," what he was really saying was that the only thing he could not doubt was that he was thinking at all. What would happen, the yogis pondered, if they doubted that thinking was primary? The yogis, therefore, said, "I am, therefore I think." They sought to look beneath the mind, under the hood of the car, so to speak.

The reductionists taught us to examine the parts to understand the whole, and that is much of what science does. When we examine the parts, we can learn a tremendous amount about each distinct thing—how an individual liver cell works, or how a muscle contracts. But a liver cell does not exist independently of the entire organ, which cannot exist independently of the body surrounding it; and a muscle cannot contract without bones, tendons, ligaments, and a nervous system to support and engage it. In the same way, we cannot exist, and do not exist, separate from the world we live in: this planet, the air, the water, and the land and the food that grows from it. We are a bundle of processes. The mind is the part of that process that coordinates incoming information, stores information, and directs us to act on that information in particular ways, depending on how that information, or contact, makes us feel. If something feels safe and secure, we move toward it. If it feels dangerous or threatening, we move away.

Yoga, first and foremost, is a practice for our mind, even though sometimes we think of it mainly as postures for our bodies. As we have discussed thus far, the mind and the body are considered to be not distinct and separate but a description of two integrated and interdependent processes that have been separated from each other only for the sake of examination, and in order to understand their various functions. In actuality, they are not separate. As a unified field, the mind-body complex has both seen and unseen aspects, the body being what we can see, and the mind, emotions, memory, intellect, and feelings being representative of things that we can feel or sense, but not see. The seen enlivens the unseen, and vice versa. For example, the sensation of love or fear is sometimes first felt as a knot in our stomachs, or in the quickened beat of our heart, before we identify it as a thought or emotion. In the same way, our posture, either slouched or standing upright, sends a signal to our brain that makes us feel that we are either avoiding, withdrawing from, or engaging with the world. We can change our attitude and our mood simply by changing our posture: our bodies reflect mental attitudes, and our minds are reflected through our posture. Look at any high school student slouching in their chair to avoid being called on when they haven't done their homework, and you have an obvious example of the mind's wish not to be seen reflected through posture. Examples abound, because the body is where we see the inner being of people reflected—what they are thinking, what they have been through, what they are feeling.

Since yoga is so largely thought of as a practice of postures, it is only natural to question how it is that by placing my body in a particular position my mind becomes balanced. Postures are physical things—why should they change the way the mind functions, or change what we can call, for lack of a better term, mental content? Isn't this an indicator that my mind and body are not separate, but interconnected? The answer would seem to be yes. The effects of yoga via our physical bodies are nothing short of amazing. Recent studies have shown that a regular practice of yoga—even in modest amounts—may help lower blood pressure; reduce anxiety and depression; reduce back pain; reduce symptoms of PTSD; reduce occurrences of asthma attacks; improve posture, strength, and coordination; improve breathing and cardiovascular health; support self-regulation in children; and so on. It's remarkable that one practice can have such a wide range of effects—yet it does. Many of the illnesses or problems that yoga helps, such as high blood pressure, depression, PTSD, and self-regulation, are mental and emotional impairments that can be brought about by traumatic incidences or an excess of stress in our lives. These events that happen to the body immediately affect what we call our mind, but they also get imprimpted within our physical body. Somehow physical practices such as yoga asanas, done with awareness and intention, can begin to release the hold that traumatic events have on us, by releasing them first from the body. Within medical science it is known that diseases such as certain cardiovascular problems (for instance, hypertension), cancer, adult-onset diabetes, and irritable bowel syndrome can be brought on by long-term low levels of stress and inflammation (excess stress, on its own, also leads to inflammation), which is a clear indicator that the stress of life can lead to the breakdown of our bodily functions.

When our body senses stress, it releases cortisol and adrenaline, which are neurotransmitters and hormones that both fight inflammation and give us energy, and are released when we do even simple things like getting up out of a chair to walk across the room. They are released when the brain senses activity; they are not the "bad guys" of the neurotransmitters. They are also released when the brain is told that the body needs to respond to an environmental load or demand that could be stressful—a late bill, an overdue homework assignment, a traffic cop, a fight, or a virus that needs to be attacked, for example. These stressful situations dissipate when we have dealt with the problem, and as the stress dissipates, the hormones are removed from the bloodstream, reabsorbed back into our cells, or passed out through urine.

However, when stress is perceived at a constant or repeated level, the body will keep producing them at a rate faster than they can be removed, which leads to an accumulation of stress hormones in our bloodstream. At that point, the opposite begins to happen, and cortisol, which helps fight infections, collects in the bloodstream and becomes toxic, and inflammation begins to increase. This is how low levels of stress, over long periods of time, lead to inflammation in the body. The ability to move inflammatory markers from the blood are strong factors in determining whether or not you will develop one of those diseases caused by stress or inflammation. Low and or high levels of stress are not completely a problem of environment; they are an indication of our mind's ability to handle stress.[3]

Alia J. Crum's conception "stress mindset" (an offshoot of Carol Dweck's "growth mindset") is another example of this: stress is not inherently bad, but it is our perception of stress that will determine our response to it, and how it affects our health, emotions, and resiliency. While increased inflammation is a by-product of stress, there are other chemical reactions that accompany how our mind deals with tension. One example of this is that we eat poorly because our body under stress does not produce all of the digestive enzymes it needs to break down complex foods. The occupational therapist Anne Buckley-Reen shared with me that when we are under stress the only digestive enzymes we produce are those that can digest things like bread, pasta, and sugar—comfort foods. If you have ever wondered why, when you are feeling overwhelmed, all you want to eat is a pint of ice cream or a donut, this is the reason. Our sleep becomes disturbed because of the excess adrenaline and cortisol in our bloodstream, and so on. One reason that we feel better after we do some yoga practice is that exercise is one of the primary things that will help remove excess cortisol and adrenaline from the bloodstream. Any exercise, in fact, will do this, but yoga has the added benefit of allowing us to access our nervous system—which perceives stress—through our breathing, and direct it toward perceiving balance, safety, and steadiness.

Many of the imbalances we experience arise when our body, nervous system, and mind are not cooperating with each other, when one of the systems dominates the others. For example, we often ignore hunger, thirst, or fatigue because we are focused on completing some project. Our body is sending signals that we should take a break, but our mind says, Do more. The nervous system gets caught in the middle and on many occasions obeys the dictate of the mind, at the peril of the body, and slowly other systems suffer. When we ignore fatigue and push ourselves too hard, our mental faculties become impaired, our immune system weakens, our strength decreases, and even our internal organs can become stressed. This happens from the general stress of everyday life, or from traumatic experiences that cause us to become disconnected.

One of the meanings of *yoga* is "union"—that special time when all things come together. When we bring our body, nervous system (through breathing), and mind together at the same time, and they begin to cooperate, interesting things begin to happen. We release stress, we heal old traumas, and we prevent ourselves from collecting new ones. We learn how to harness our ability to choose wisely, and override compulsive tendencies of the mind to overdo it—or to underdo it. The underlying premise of practicing yoga is quite straightforward: by moving, breathing, and concentrating in a particular way, we will bring our body, nervous system, and mind into a coherent, cooperative state.

This brings us back to Vyaas's point, that yoga is both a practice and a state of concentration, or rather, it is the practice of developing the type of mind that reflects awareness and is therefore steady, open, aware, at ease. This is a trait of mind. As mentioned earlier, a trait is different from a state. A state will change; it is something that occurs at a certain time. A trait, however, is the underlying character of something. In the practice of yoga, we are trying to change the underlying trait of the mind, so that its base, its support, is clear, calm, and adaptable.

When the underlying character of our mind is transformed, then the things that come our way during the day cease to be stressful, because distress occurs when our mind is not at ease, and not because of external conditions. When we try to pay attention, or try to concentrate, we can do so for short periods of time, but then our mind wanders away from whatever we are trying to focus on, and when we attempt to force it back, it doesn't stay. We've all experienced this. We can experience profound states of consciousness, but then somehow when we come back down from them, we have the same problems as before: anxiety, anger, competition, arrogance, you name it.

There are also times that we can get into the "zone," and have periods of great immersion in an activity, whether sports or a creative endeavor—but those times when we are in the zone have to come to an end, because being in the zone is also a mental state. The development of a positive mental trait is much more enduring then fleeting states, and can be practiced in daily activities such as listening to our partner or children. When we are fully and completely present to that which is in front of us, we are unified in body, breath, mind, and spirit. This brings connection, satisfaction, joy, and a feeling of purpose to our lives. The development of this kind of internal strength was spoken of twenty-five hundred years ago in the *Mundaka*

Upanishad: "Naayamatama balahinena labhyah"—the inner self cannot be grasped without strength.[4] The strength they were speaking of is internal. It is the strength of the mind and the strength of devotion, and not just the strength of the body.

IS MY MIND IN MY BODY?

Our bodies are the homes in which we live, and are the physical expressions of our mind, nervous system, emotions, karma, and potential. Even though yoga is primarily about the mind, it's impossible to see our mind. Where is it exactly? What is it composed of? We have no idea. Literally thousands of books have been written about the mind over the past several centuries, yet there is very little agreement on what the mind actually is, or where it is located. The answer that yoga gives is that the mind is a neutral field in which experiences occur or are perceived. It can expand and contract to fill any space, however big or small; for example, it can expand to contemplate the nature of the universe, with its billions of galaxies and trillions of stars, or it can contract to focus on a single atom, a single cell, or the lightest sensation of touch where the breath leaves the tip of the nostrils. The mind, when it has content, is composed of thoughts. When there are no thoughts, the mind is a neutral field.

The ancient yogis presented the idea that we have not just one body but three. The first is our physical body, made up of the food we eat and the water we drink. The physical body can be strengthened, stretched, molded, and changed through diet, exercise, and environment. Our physical body is not limited to just our own personal limbs; we have an extended body, as Dr. Deepak Chopra has spoken about in *The Seven Spiritual Laws of Yoga*, and that is the world we live in. Our physical body does not, and cannot, exist as separate from the environment. We depend on air to breathe, light to see, fire to cook, sunlight to grow, ground to walk on, space to inhabit; without the biosphere, we do not exist as a physical body at all. In fact, some say that our physical body has developed in response to the environment or, perhaps, as part of the environment. The idea of the world around us being our extended body is, to me, one of the most profound and obvious reasons for practicing the yamas, and behaviors like non-harm and kindness. Not only are the elements and people around us extensions of us, but we are extensions of them as well. There is truly no "other" out there; there is only all of us, together, as one, huge, simultaneous happening. If we could inhabit that level of consciousness as an integrated part of our lives, we would truly live in unity consciousness.

The second body is called the subtle body, and it is made up of the breath, the mind, and the intellect, called *prana*, *manas*, and *vijnana*, also called *buddhi*. It is called subtle because we cannot see our breath, our mind, or our intellect. But we can feel them and sense them. Our mind, according to the yogis, is the seat of thought, emotions, desire, and memory, which differs from our intellect, which is the faculty of discrimination. While in our mind we might desire, say, a donut, our intellect will be the one who says, "Maybe have some raspberries instead." Our intellect lies closer to our sense of self. Dr. Daniel Siegel has said that the mental processes of awareness, which the yogis identified as the subtle body, can be used to meditate on our entire sphere of experience, and then to be able to distinguish the experiencer from the experience that is being had. He calls this the Wheel of Awareness, a straightforward and profound practice that guides you through meditating on your five senses; the interior sense of your body (interoception), which he calls the sixth sense; your mental activities, the seventh sense; and your sense of interconnectedness, the eighth sense. These senses exist as the rim of the wheel, where our senses meet experience. The hub of the wheel, however, is our clear, calm, receptive, open, and aware sense of being. It is our center, a center that is connected to all other centers, and perhaps where we can sense divinity within us.[5]

The *Chandogya Upanishad* has a similar description of our lives. It says that the heart is our hub, and the rim of the wheel is the wheel of life. The spokes of the wheel are where we get lost in karma, false identification, and experience. For example, sometimes when we start a yoga practice, we immediately feel that we are in touch with a part of ourselves that is open, aware, free, and loving. We keep going to classes, and one day we notice that the person next to us is more flexible than we are, and doing a posture really well. We think, "Wow, I want to be able to do that. When I can put my head on my knees I'll really be making progress. Maybe I'll take a weekend workshop to learn some tips." So we sign up for a workshop, but then realize that we need a better yoga mat because with a good mat, maybe your yoga is better. And perhaps a nice yoga outfit, in case we meet someone who could be a potential partner. After all, yoga retreats are for potentially finding your perfect yoga soul mate, right? And of course, now that I have the right outfit, maybe some mala beads to complement the outfit—those will look very spiritual. Then with the new look solidified, and my hamstrings finally opening up, I realize I've gone from open, loving awareness to creating an entirely new false identity. This is how we get lost in the spokes, and it happens with many of our different undertakings. But the good thing about spokes is that they go in two directions—away from the hub, and also back toward it. At a certain point, we can recognize when we have gone off track, or when we have gotten lost, and then use that recognition to travel back to the heart, to the hub of our being.

Our third body is the body of bliss, the bliss of contact with knowing who we really are—our potential, our infinite creativity, our source of being as pure consciousness. Bliss here is not the bliss of a fleeting indulgence, but the unlimited joy of being; something we have all felt when we experience an uncaused joy—when for no reason, we feel inexplicable happiness and contentment, and everything feels okay. The bliss body shines through the other two, but it becomes more and more dim as the outer bodies are thicker and grosser and allow less light to shine through, until we actively work on making ourselves more porous, less shell-like. The bliss body can be seen in the shining eyes of people like the Dalai Lama, Desmond Tutu, and maybe the other living saints of our times, as well as yogis and meditation practitioners, and especially in the eyes and smiles of loving, caring, thoughtful people.

Another name for the body of bliss is the causal body. The word for causal is *karana*, which comes from the verbal root *kr* (where the word *karma* comes from), which means "action." All actions that we perform leave an impression, called *samskara*. If you eat ice cream for the first time, and really enjoy it, the impression left in your mind is a positive experience of ice cream. Eating is the action, and the impression, or memory, is the samskara. Along with samskara there is a subtle desire called *vasana* (which means "fragrance"), which goes hand in hand with the samskara. If the experience was positive, then you'll have a desire to repeat the experience. If it was negative, then you'll have the desire not to repeat the experience. The place where samskara and vasana are stored is the causal body. They are the cause below our intellect, mind, nervous system, and body for the likes and dislikes we have, and the reasons why we do the things that we do, or have the quirks that we all have. They are the billions of fragrances that make up our individuality, and are the underlying cause of the talents and obstacles we have in our lives. Karma is action, samskara is memory, and vasana is the desire that arises from memory.

The cycle of action, memory, and desire. The yogis uprooted this cycle by not acting upon every desire that arose in their mind.

Every action that we perform needs to resolve itself at some point, or in the words of Newton, every action has an equal and opposite reaction. If you throw a ball up in the air, eventually it will fall back down. The law of karma says that any action we perform will be reciprocated at some point. If I do something nice for you, eventually you'll do something nice back to me. But what happens if I do something nice for you, but then I die before you can reciprocate? The karmic bond created by our interaction doesn't dissolve because I've died; it will just get taken care of at a later date. All of our unresolved actions, and impressions of those actions, are stored in the causal body, so that when we die these residuals will eventually take on a new form to find resolution to our karmic debts. This is, in a truncated form, the theory of transmigration, or the cycle of birth and rebirth. It is an important idea in the three-body scheme, because it is in the causal body that our reasons for being born are held, our purposes in life. Within our three bodies, experience is knotted to an idea of an individual self, to who I think I am and what my story is. Mystical practices untie these knots so that we can experience unfettered freedom. Beyond the three bodies, according to the yogis and mystics, is pure existence, unlimited being, and non-local consciousness. But this is not an easy experience to have; sometimes we have a spontaneous but short-lived experience of pure consciousness. Even a short-lived experience like that can be transformative, but to have freedom as a continuous experience comes from either grace or practice, depending on whom you talk to.

Yoga as a practice was designed explicitly to address, strengthen, and purify all three bodies and to move us toward a continuous experience of freedom. Through postures and clean eating, we purify the physical body; through breathing, meditation, and chanting (to be covered later), we purify the subtle body; and through service and thinking of others first, we purify the body of bliss. And within this model, where is the mind? It exists in the subtle body, as a vehicle of thought. It is not the highest aspect of our inner world, just one that reflects images, ideas, and experiences that are put before it. What we do with those thoughts, images, ideas, and experiences is the jurisdiction of the intellect, or the buddhi. Buddhi is the source of willpower and discrimination. The ability to choose where we want the mind to focus has its basis in buddhi, which means that mental focus comes not from the mind but from something deeper. Forming a relationship with this deeper aspect of our unseen world is empowering and freeing. Loosening our identification with the content of the mind, namely, our thoughts, is an important first step in moving to deeper levels of who we are.

The most important tool we have in the process of moving inward is our breath. In the model of the three bodies, the sheath that lies between the body and the mind is the breath; it is the link to the inner levels of the mind, and the intellect. If you want to work with calming the mind, the breath is a failsafe place to begin. The mind is important because we spend so much of our waking hours in it—thinking, planning, remembering, feeling, absorbing information—but it is not the deepest or most profound part of us; it's just a way station for incoming experience. What lies below the mind is where peace is experienced, and for that, we have two wonderful tools to lead us: breath and awareness.

To sum up, the three bodies and five sheaths are:

* The physical body and the sense organs are associated with suffering—whether it be physical or emotional pain—and comprise our most visible body

* The breath, the mind, and the intellect are unseen, but can be felt and sensed, and so are called the *subtle body*. The breath is linked to the nervous system, which is the mechanism we use to perceive and live in the world through our physical body. The mind likes to compare and measure things all the time.

* The intellect is the source of the I-sense, or identification, and is where we choose which direction to go based on our sense of identity. If we identify inwardly, we move toward knowing who we are. When we identify outwardly, we identify with objects.

* The most inner body is the body of bliss, which is our innate essence of existence.

* Pure consciousness is the infinite field of existence and being, from which all of the bodies arise as fleeting, changing flashes of experience.

So these are the three bodies and their five layers. How can we move from the physical body to the deeper levels of our being? The first step is to ask, who am I?

six

{ WHO AM I? }

ACCORDING TO THE *YOGA SUTRAS*, people come to the practice of yoga only after they have recognized, to some degree, that they are suffering, or that something is missing in their lives. Suffering is caused by obstructions in the field of consciousness. Consciousness, awareness, or self-knowledge is bright, clear, and open; knows itself; and is thus eternally present. Obstructions are like clouds that cover awareness, so that our identity rests with the clouds, and not with the pure field of knowing. The obstructions are called the *kleshas*, and there are five of them, which have been lucidly presented by Patanjali, who makes sense of our muddled minds in just a few verses. Yoga is optimistic about suffering and does not put too much emphasis on it; rather, it says that suffering can be worked on, actively, by doing three specific things. Patanali calls these the "actions in yoga," that is, the things we can do that make yoga work. They are:

1. Practices such as asanas and pranayama—*tapas*

2. Chanting and study of the philosophical texts— *svadhyaya*

3. Surrender to God, the Divine, or the unknown— *Ishvara pranidhana*

We'll go into each of these in depth later, but let's start with the obstructions, the kleshas. Once you understand a bit about the obstructions, it helps to define what exactly it is that we are working to reduce, or as Patanjali says, to thin, when we do yoga.

The five kleshas are:

1. *Avidya*—not knowing who we truly are (often translated as "ignorance")

2. *Asmita*—creating a story or narrative about who we think we are

3. *Raga*—attachment to things that are pleasant

4. *Dvesha*—attachment to things that are unpleasant (often translated as "aversion")

5. *Abhinivesha*—fear of extinction (often translated as "fear of death")

AVIDYA

The first of the five kleshas is the field within which all others grow, and it is, very simply put, not knowing who you are, called avidya. When you know who you are, all of the other obstructions fall away. Knowing who we are is *vidya*, or "knowledge." It is the experience of ourselves as awareness, consciousness, or the witness. When we know who we are, it automatically uproots the other types of suffering. The syllable *a* in Sanskrit can indicate the opposite or absence of something. If *vidya* is knowledge, then *avidya* is an absence of knowledge. We might have lots of other types of knowledge: school and book learning, expertise in art or music, science or math; we might know a lot about politics or medicine, but what we don't know is our essential nature. Avidya is more often than not translated as ignorance, but it is not ignorance about everything; it doesn't mean that we are completely and utterly idiotic. It's ignorance of only one thing, of knowing who we are, which is actually the most important thing for us to know. When we know who we are, the false identifications we have fall away like leaves from a tree in the autumn, and what remains is consciousness.

ASMITA

When we don't know who we are, we have to create an identity, a narrative, which is essentially false in that this assumed identity can change, depending on what we are drawn to, or find unappealing. This is called asmita, or "I-ness." If we don't know who we are, then we have to make something up. The created self-narrative of asmita does not mean "ego"; it means the stories we construct that take the place of self-knowledge. It's a narrative that shifts throughout our lives, eventually, perhaps, leading us to a narrative that is free from storytelling, and becoming a narrative of simply being present.

When I was a teenager, for example, music defined my narrative. When I identified as a punk rocker, I had a Mohawk and wore ripped jeans, a leather jacket, safety pins everywhere, and T-shirts with my favorite bands on them to announce my affiliation. A few years later, I discovered the goth scene, dyed my hair black, got a few more piercings, and exclusively wore black. One narrative followed another, until I discovered a world that had an infinitely healthier story. However, like each narrative that I entered into, the identities that we all enter into have the assumption woven into them that what we are aligning ourselves with are real identites. But they are not; they are temporarily assumed identities that fill the hole of not knowing who we truly are.

Around 1986, after I graduated from high school, I was working in a record store called Bleecker Bob's in Greenwich Village, New York City. A guy working there named Ted was a vegetarian; he had done yoga in the 1970s with the yogi Amrit

Desai, read a lot of philosophy books, and sold MDMA on the side. I was living a completely unhealthy lifestyle at the time, living on Big Macs, pizza, cappuccinos, soda, beer, tequila, and cigarettes. That was my daily diet. I only ate a vegetable if it was tomato sauce on pizza or the lettuce, pickle, and tomato in a Big Mac. Ted told me about vegetarianism, and I thought, "This sounds really good. I think I want to be healthy." So the next day I started. For the first month, I didn't know what to eat, so I mainly ate iceberg lettuce, apples, and rice cakes. Little by little I learned how to cook some macrobiotic dishes from a book I picked up called *The Macrobiotic Way*, by Michio Kushi. In that book, there were some exercises that Kushi said you could do to open the meridians and keep energy moving in your body, so I tried them, not knowing that they were essentially yoga poses. This was how I got into yoga.

Very quickly, my clothing changed from black to colors, my hair grew (and, at a hefty price tag, I dyed it back as close as possible to my original red), my music taste changed, and my lifestyle radically shifted. Instead of coming home at five or six o'clock in the morning, I was getting up at five or six and chanting or doing some meditation. It was a new identity, a new narrative, but it was a narrative that I thought would bring me closer to understanding "Who am I?" The problem with narratives, though, is that no matter how well intended they are, if you hold too fast to them, they will limit you, no matter how based on freedom they are. Thinking that I was a yogi, or spiritual, became another false identity, one that allowed me to think that I knew better than other people who were not eating a vegetarian diet, doing yoga, or meditating. Any narrative is a bind, any narrative is illusory, any narrative will keep us stuck in avidya. The only narrative that leads to freedom that we can tell ourselves is one of no story. If during the day we can spend some time watching every story that comes up in our minds, and not identify with the story, the mind will get very quiet. Thoughts can't dwell in a mind that doesn't indulge them, and thoughts are stories—and of course the hardest story to tell ourselves is that we are not our bodies, because most of our false narratives stem from identification with our bodies.

The great epics of the Hindu tradition that pass down spiritual teachings are long, elaborate, intertwined, and filled with complicated stories rife with moral dilemmas and villains who often are quite good at heart. So it's not that stories are bad; it's that if we are not cognizant of how we create them, we will be bound by them. The clinging to our stories can be either like living in a haze, or denying outright our true self, which on some level will also make us feel that we are not living in alignment with our highest self. James Baldwin recognized the pain of this conflict, and described the problem of avidya very clearly: "This collision between one's image of oneself and what one actually is is always very painful and there are two things you can do about it, you can meet the collision head-on and try and become what you really are or you can retreat and try to remain what you thought you were, which is a fantasy, in which you will certainly perish."[1] Dr. Robert Svoboda, though, has another view on asmita, which is softer and very practical. He said: "It is essential to have a self-narrative that is healthy, and a critical part of a healthy self-narrative is the understanding of what it is that gives life meaning for you, and what you are doing to actualize that meaning."[2]

Asmita, in this sense, can be a way of discovering through personal storytelling what our purpose in life is, and if we are living in accordance with it. The neurologist and psychiatrist Viktor Frankl, who was also a Holocaust survivor, discusses a similar view on developing a healthy narrative in his multimillion-copy-selling book *Man's Search for Meaning*. However, his twist is not that there is a meaning *of* life, but that life demands something of us, and if we respond to it appropriately we discover meaning *in* our own lives. He says, "One should not search for an abstract meaning of life ... Man should not ask what the meaning of his life is, but rather he should recognize that it is he who is asked. In a word, a man is questioned by his life, and he can only answer to life by answering for his own life ... The meaning of life changes, but it never ceases to be." This is indeed a very healthy approach to asmita, to developing a healthy self-narrative that is not self-centered, but is responsive to the world and meaning within it as it changes.

RAGA AND DVESHA

Our stories fall into two basic categories: things we like, and things we don't like.

As the Third Zen Patriarch said in his *Verses on the Faith Mind*,

> *The Great Way is not difficult*
> *for those who have no preferences.*

When love and hate are both absent
everything becomes clear and undisguised.
Make the smallest distinction, however,
and heaven and earth are set infinitely apart.
If you wish to see the truth
then hold no opinions for or against anything.
To set up what you like against what you dislike
is the disease of the mind.

The disease of the mind is avidya, not knowing who we are; our opinions are asmita, our false narrative; and asmita is defined by our likes and dislikes: raga and dvesha.

Both raga and dvesha are attachments. When we like something, we are attached to liking it, and when we dislike something, we are also attached—to not liking that thing. I like coffee, I don't really like green tea; I like yoga, I really don't like boxing. If one thing fills us with pleasure, another thing perhaps fills us with disgust or a feeling of superiority for not liking it. Sometimes we attempt to soften the dislike of something by saying, "That song/food/person is not for me . . ."; but at the same time, we are also implying that something else indeed is for me. We often think that the things we like are the things that we are attached to, but the things that we don't like, and that we are dead set on continuing to not like, are also attachments. In fact, sometimes the dislike of something or someone can actually give us great pleasure! So don't think of pleasure as good and aversion as bad; think of them both as attachments to ways in which we define a false sense of our individuality and solidify our narrative.

Another option is to decide to become equally and unequivocally attached to everything and everybody. Once when I was visiting a philosopher friend in South India, Perumal Raju, I said in passing something about non-attachment, trying to be philosophical. He laughed and said, "I don't believe in non-attachment, I prefer to be equally attached to everything I see in God's universe, good and bad." I was struck by this statement, which was opposite to everything I had heard in the yoga world so far. He seemed to be saying, as the Third Zen Patriarch did, "When love and hate are both absent, everything becomes clear and undisguised."

ABHINIVESHA

The fifth reason why our true self is hidden from us is fear. Fear is rooted in asmita, because not knowing who we are makes us hold on tightly to our story, like a person lost at sea clinging to a log or a leaky raft. We think somewhere inside of ourselves, "If we are not our story, then who are we?" Often *abhinivesha* is translated as "fear of death," but Swami Hariharananda argues that death is not the ultimate fear, because life and death are an integrated, natural cycle in Hinduism. Death is not final, because, as mentioned earlier, all of our unresolved actions, in the guise of samskaras, will take on a new form to move toward their resolution. Nothing, then, is really ever born, and nothing dies; it's just nature changing forms. It is not really death that we fear, then, but what death signifies to us, which is the annihilation of our sense of identity. It is extinction that we fear, and not the cycle of life and death. Life and death are a continuum, so there is nothing to fear, because birth and death, coming and going, are eternally linked. But extinction is permanent. There is no coming back.

Since we all know with some degree of certainty that we exist, then the removal of that knowing pulls the rug out from underneath our feet. If we do not exist, who are we, where are we, and what was all of this? A giant dream? A huge illusion? If that's all it was, then what was the point of feeling and experiencing so many things? Love, hate, pain, pleasure, fear, attachment, jealousy, compassion, longing, hope, certainty, confusion. If all of those things are just a dream, a grand illusion, what's the point? These are important questions to ask and to ponder. But where exactly are we experiencing all of these feelings, thoughts, and emotions?

We experience them in the realm of our mind, where thinking happens; and all of our thoughts are based on past experience, on exposure to experiences. The essential point of experience, according to the *Yoga Sutras*, is to remind us that we are aware. The world, according to Patanjali, exists for knowledge and liberation. This means that our awareness can go outward into the world to gain knowledge and experience, and inward toward our sense of self to "attain" liberation. The problem is that when we imagine our experiences as absolutely real in the sense of an unchanging and true reality, then our awareness gets lost in the stories we are telling ourselves. When we use our awareness to see that the experiences and stories are always changing, then awareness looks back in on itself, and past the screen of the mind upon which all stories are projected, to see itself as the witness of the stories.

Yoga is essentially a practice of looking into ourselves. Avidya and asmita are at their strongest when we do not look into ourselves to see the transitory nature of our created narrative. The narrative and innate sense of "I am" become as if one, and then all we really know is our story! Therefore, if I don't know who I am, then I cling to my story, and fear exists within that clinging because at that moment I am the sum total of my story, and if my story evaporates, then I will cease to exist. That is what is known as extinction. There is no rebirth, there is no continuum, there is just total annihilation. That idea, indeed, is scary. However, if we have learned to become inwardly aware, then when the story begins to loosen its hold, the "I am" sense has already been strengthened, so fear doesn't have a foundation to stand on. What remains is wholeness, peace, and ultimate knowing. There is no fear—not even a trace of it—in pure "I-ness," because existence existed before there was even a concept or thought of fear.

Scientists, philosophers, and yogis of all types throughout the ages have sought to find that screen where all experiences are projected, and also the place in our minds where all memories are held. Neuroscientists say that our thoughts and memories are the work of the synaptic firings of the brain. The yogis have said that looking with the eyes of logic will never allow us to see deep enough, because logic is concerned with measurement, logic is another story within the story we are already telling about "trying to figure it all out," but the screen that all measurements are projected on is consciousness, which is without measurement. It is existence, unlimited. It is existence without any additives. It was never born, it never dies; existence just is. That is-ness is the part of us that is truly who we are, it is our sense of "I exist," of "I am," it is filled with purpose, filled with its own meaning, and its meaning is expressed through this majestic manifestation that we are all an integral part of. That feeling can never be destroyed; it is always there within us. It is the feeling that Viktor Frankl and others held on to in the Holocaust and that allowed them to survive. It is the feeling of compassion that the Tibetan monks who were imprisoned by their Chinese captors held on to in order not to descend into hatred and the cycle of hostility toward their captors, but to instead still feel compassion for them; it is the feeling that anyone who is captured, imprisoned, or traumatized holds on to in order to maintain their dignity; it is existence. It is not just who we are individually; it is all of us, together, simultaneously.

The artist Francesco Clemente summed up the kleshas in a Buddhist framework and poetic way that I find appealing: "We don't know where ignorance began, but there are tools. There is an education that can bring ignorance to an end. The reason for ignorance is desire. There is an education that can free you from desire. To be free from desire means to be free of the fear of death, and the condition of people who are educated in that way is a condition of freedom. This is basic."[3]

The five kleshas are the obstructions to inner knowledge. They have a remedy, an upaya, which is called kriya yoga, which are the three practices mentioned at the start of this chapter: tapas, the practice of yoga postures, pranayama, and meditation; svadhyaya, the chanting or study of texts and scripture; and Ishvara pranidhana, surrender to God for those who are theistic, and to the unknown for those who are not. These are the tools that Clemente referred to above.

These three practices thin the obstructions that cloud the field of consciousness, so that the light of awareness shines through with greater strength and clarity. However, they are called indirect practices because they do not work directly on the root of thoughts, but instead create and support particular behaviors that create patterns of thinking that are conducive to awakening. In the next chapter, we will discuss both the yamas and niyamas, where these three actions appear, and the actions themselves, in more detail.

{ THE FIRST TWO LIMBS }

ALTHOUGH THE FIRST LIMB OF YOGA is called yama, which means "restraint," we began our discussion of yoga with the third limb, asanas, because the body is the easiest place to start a spiritual practice. We can see it, we can change its shape, see quick gains in strength and flexibility, and improve the way it functions—for instance, our digestion or how well we sleep. Changing the shape of the body automatically begins to change the shape of our mind, because mind and body are a continuum.

A simple thing like being able to touch our toes, when before we could not, affects how we feel about our potential. Possibilities begin to expand. Expanded capability of the body and expanded sense of our potential for growth and change go hand in hand. That feeling is exhilarating and encourages us to continue, and to do more.

Along with this, the character of our thinking processes naturally begins to change. Perhaps we get angry less often, or develop a little bit of patience. Maybe we feel a little more forgiving, or are able to relax our need to control everything and everyone. For some, it may be experienced as a feeling of contentment, a letting go of striving so hard as we learn what it feels like to be more in the moment. However, as with our bodies, there is a threshold for this. We get to a point where our hamstrings or hips are not opening naturally on their own, or our patience starts to slip again. We have to apply extra effort to get something to move. Yoga practice can quickly have some beneficial effects on our body, mind, and ability to handle stress, but then it will level off, and we see that we have to apply a little more effort.

The initial effects of practice are like picking low-hanging fruit. Within the first few weeks or months we see change. But then we have to stick with it over many years to get to the deeper levels of our habitual patterns. To uproot recurrent behavior patterns takes work and can be frustrating at times. We can wonder things like, Why do I still lose my patience over stupid things? Why does jealousy flare up in me? Why do I still get easily frustrated with people and situations? As we get used to paying attention and become sensitive to what is happening in our minds, we can develop the ability to step back and take a look at ourselves, and identify what the causes might be. Perhaps you lost your patience because you were tired, or overworked, or hungry. Or perhaps it was because someone did not do something exactly the way you like it, or because you were not listened to. These are the things that we can come to observe about ourselves. Yoga and meditation practice are meant to show us clearly how we are behaving and reacting and provide us with options for being responsive in a more constructive way.

We usually realize after it is too late that we have reacted wrongly, or more forcefully than necessary, in a situation that triggers our weakness (like impatience). Can we catch those moments before they happen? The musician Laurie Anderson related a story to me recently about her partner, the late musician Lou Reed, who was a very loving person but had a quick and sometimes ferocious temper. Toward the end of his life, when his liver was failing, he began to catch more quickly the moments when he'd lose his temper, and apologize right away. Then, the outburst and apologies would come exactly at the same time; finally, the apologies would come first: Lou would say, "I am sorry, I almost got mad." It is an exquisite and fine attention to our mind that enables a journey of self-awareness to unfold like this.

When we are ready to start working deeper, the yamas and the second limb of niyamas (there are five subcategories of each), become the typical next steps of a yoga practice. They are the boundaries we purposefully set for ourselves in regard to our personal behavior. Healthy, purposeful boundaries help us know where we stand. When we don't know where we stand, we can end up allowing ourselves to get into situations that are not beneficial for us. In spiritual practice, freedom starts with restriction. In America, we think that freedom means the ability to do whatever we want, whenever we want, wherever we want. This is not freedom: it's hedonism. Freedom in yoga is the experience of a self, a consciousness that is not dependent on external actions or circumstances, but is an internal, uncaused happiness. It is the happiness of knowing who you are at all times and in all circumstances.

In the course of studying the Torah with my rabbi, Mendel Jacobson, he and I discussed time, birth, and death. Mendel said that, according to the Torah, we aren't born once, but are constantly being born, and that constant birth-and-death is God's restraint on us so that our vision remains fresh and new. Otherwise, we would become robotic. In Hinduism, the cycle of birth and death is called samsara, the net of illusory existence. Karma, or our own actions, can keep us ensnared in the net or free us from it.

Rabbi Mendel made a beautiful analogy. He said that according to the Kabbalah, rain is described as restraint. Restraint breaks water up into raindrops; if there were no restraint, rain would be one huge drop, a deluge, and endless wall of water, and the universe would be drowned in it. God's restraint allows for diverse manifestation and, because of this diversity, a freshness and openness of mind. Without the freshness of every new day, we live in a box, which is a kind of hell; but to live in no box is to live in freedom. Restraint is necessary for existence because it leads to freedom.

The five yamas define the yogic idea of boundaries for us. The yamas refer to a behavioral code that is based on personal responsibility and how we apply ourselves to our relationships. When our minds and hearts are aligned with consciousness, then the yamas, as Deepak Chopra has said, are spontaneous displays of elevated, or enlightened, behavior. They become natural expressions of truth. But until they become natural, we can try to do them as a practice. Just as we try to improve the functioning of our bodies in yoga postures, we can try to improve our interactions with people in the world through the yamas.

The first yama is called ahimsa. As mentioned earlier, in Sanskrit the letter *a* before certain words means "the opposite" or "an absence." *Himsa* means "harm," so *ahimsa* is often translated as "non-violence," which is the opposite of harm. On a deeper level it means that there is an absence of harm within us; there is no inclination for it in us, no chance that we would cause harm to someone. When this level of kindness is embodied in us, a person who comes in our presence feels peaceful; even if someone else is present whom that person happens to hate, the hostility will cease.[1] To have a complete and utter absence of harm in our hearts and minds is truly freedom and love for all beings.

But if we are not at that level, we can practice non-harm in other ways. We can eat a diet that is less harmful to animals; we can make choices in our lives about the resources we use. We can pick one person who makes us upset and for whom we wish bad things, and practice directing kind thoughts toward him or her. Or, for an hour a day, we can practice ahimsa by paying careful attention to our minds, thoughts, actions, and behavior. Even in that short amount of time, we can learn a lot about ourselves. Ahimsa as an absence of harm also can mean kindness. When we do not have harmful thoughts toward ourselves or others, we have kindness and compassion instead.

Reactivity is a driving force in our lives, and the ahimsa is the guideline for how we can treat ourselves, the people around us, and the world without harm. We cause harm on a daily basis, usually in unconscious ways. If we are unhappy with the way we look, we can follow harmful dieting habits, exercise too extremely, and experience shame and frustration. In our interactions with people around us, we may experience judgment, criticism, envy, or competition; we might even wish them ill. Sometimes we act harmfully or are untruthful in order to protect ourselves. On occasion we use unnecessarily harsh words that upset our own minds and those whom we direct them toward. Ahimsa is the first yama because it is the most important one; it is the basis for a calm mind and loving heart. The seeds of all of the other yamas have non-harm as their base, similar to how the other kleshas grow in the field of avidya.

Satya is the second yama. It means "honesty." Satya can refer to speaking the truth, but truth said in a sweet way, one that does not cause harm or pain. This is how yogic conception of truth. In the West, we quite often prefer the "brutal truth"— somehow that does not always seem to help. The *Manusmriti*, the earliest Hindu text of laws, codes, and instructions, says: "Speak what is true; speak the truth in a sweet way; do not say what is true but not sweet, nor say what is sweet but not true. This is the eternal dharma."[2] Speaking the truth in and of itself is not enough. The truth we speak needs to be in alignment with dharma, our collective good, and not just the individual good. Satya is a challenging yama, because how do we know what truly is good for the collective? A good way to gain some perspective when in doubt about satya is to refer back to ahimsa. Ask yourself, Is this truth I am about to tell going to cause or prevent the least amount of harm, even if it is a hard truth? The most important way to tell the truth is in a way that can truly be heard. If we speak one thing and a person hears another, that can lead to more upset. So part of telling the truth, part of honesty, is saying what you need to say in such a way that your words convey your feelings, and that the person or people listening understand what it is that you are saying. In this way, according to the *Yoga Sutras*, your words will take on the power of becoming true—but Hariharananda adds, "Yogins, however, do not entertain fruitless resolutions beyond the reach of their power." This means not only speaking the truth, but letting discrimination guide your words,[3] and on some occasions, it is better not to speak at all than to speak what cannot be heard.

The third yama, *asteya*, literally means "non-stealing," but in practice it means the contentment that comes from not needing to take what belongs to others. *Steya* means "to steal," and *a*, as said before, means "an absence of," so asteya is an absence of the need to take what belongs to others. This type of stealing refers to objects that do not belong to us, as well as other people's thoughts, ideas, and dreams. It's easy to take on other people's ideas and somehow end up thinking that they are our own. The easiest way to remedy that is to credit people immediately upon using their ideas. When we have total inner contentment, then all riches are felt within us.[4]

Brahmacharya is sexual responsibility and fidelity. For monks, it can mean celibacy. For married people, it means fidelity to your partner. For single people, it means being responsible toward the intimate act of love. Brahmacharya is a huge topic in the yoga texts, because it is said that through too much intercourse, or intercourse at the wrong times, or with the wrong people, we lose our vitality. When our vitality begins to wane, so does our ability to focus.[5]

Yogis are concerned with restraining the outward movement of the senses, and sexual attraction is, to them, an impediment to inward concentration, because it is easy to get lost in the illusion of satisfaction or contentment that comes from sensory stimulation. They try to restrain that impulse, along with other things—the breath, the body, and the mind—in other ways. But brahmacharya differs from the other restraints and is on the list of interpersonal restraints (yamas) because it involves direct physical and interactive contact with other people. With the other restraints, you need only yourself. While ahimsa may involve not causing physical harm to people, it can also be harm from speech or other indirect actions, so it is a little different.

By being sexually responsible according to our stage in life or the path we are following at that time, we give the utmost respect to our partners and potential partners, and become conscious of the way we objectify people through desire. When we are not responsible with our sexuality, then the opposite occurs, and we use people for our own selfish gratification. In the texts, brahmacharya is sometimes said to be the appropriate restraint of sexual activity in word, thought, and action. It is best to decide what level of sexual responsibility you think you can honestly adhere to, so that it supports your spiritual growth and life, and does not cause you to behave hypocritically or in opposition to your highest aspirations.

Aparigraha is the confidence that comes to us when we don't desire the abilities or objects that other people have. It literally means "not grasping all around," and is sometimes translated as "non-covetousness," which sounds mildly biblical. *Graha* means "to grasp," *pari* means "all around," and *a*, as you know now, means "the opposite of." Taken altogether, it means not grasping the stuff that your senses perceive everywhere around you. As an exercise, try to look at an object, just observe it, and watch how your mind reacts to it. All objects are the creations of someone else's mind. We can use them, or not, but we give so much weight and importance to them, every day, that we lose perspective on what we need, what we want, and where our happiness comes from. If we think happiness comes from objects, we'll never be satisfied. If we think that happiness is our natural state, and can be found within our own being, then we can find happiness. Aparigraha is essentially the practice of not looking for truth or happiness in any object or person outside of ourself. Radhanath Swami put it very succinctly: "In an enlightened society, people use objects and love people; in our society today, we use people and love objects." With aparigraha, we are trying to flip this equation on its head.[6]

You can apply each of the above ideas to different aspects of your life; toward yourself and toward the other people in your life. Some of the yamas may resonate with you more than others. If you pick one small thing to work on from the yamas, like being more honest with yourself, how you express yourself, or altering your relationship with material goods, many other things can fall into place. Quite often bringing alignment to one aspect of your life opens doors to other areas that seemed closed.

There are also five niyamas, which put in order the areas where you are responsible for your interior life.

Shaucha is cleanliness. It refers to cleanliness of the mind, which means to have a friendly disposition, and to cleanliness of the body, which means to bathe and maintain the body in all of the ways it requires.[7] However, the yogi realizes that even though we clean our bodies every day, they continue to be a continual source of uncleanliness, and so we develop an aversion to our own bodies, and to contact with the bodies of others. In our modern times, I don't think we need to go quite that far, as emotional bonds through physical contact and healing touch have an important role in human development. However, the *Yoga Sutras* continue with an additional verse on cleanliness, which says that through mental cleanliness, or purification, the yogi gains purity of heart, which is accompanied by mental bliss. This sense of inner bliss leads to a condition where the mind can concentrate in such a way that the sense organs are drawn inward. From this, the buddhi becomes strengthened, and self-realization becomes possible.

The characteristics of the mind that are cleansed are called the six poisons: desire, anger, delusion, greed, pride, and envy. When these are absent from the mind and heart, an ease of spirit over comes us, and the inherent sattvic nature of the mind becomes predominant. We achieve clarity, the ability to be reflective, responsive, and happy.

Santosha is contentment. It refers to an underlying characteristic of the mind that is a happiness not dependent on external situations. To keep our minds happy and content whether in gain or loss, praise or blame, is contentment. Contentment is an even, steady mental trait, as opposed to changing mental states, which depend on fortunate circumstances for us to be happy, while we are unhappy during challenging times, or at times when we don't get what we want.[8]

The next three niyamas—tapas, svadhyaya and Ishvara pranidhana—are collectively known as kriya yoga, which are the indirect actions that lead to the thinning of the kleshas. The most direct way of removing the veils of ignorance is direct experience of one's inner being, which occurs in samadhi. For those who are not yet ready for that level of experience, we can practice the indirect methods, which will prepare us to eventually have a direct experience of the self.

Tapas is discipline, the discipline of having some type of a practice that challenges you, to a certain degree, on a daily basis. It could be meditation, asanas, pranayama, or chanting. The mild friction created through discipline encourages growth in the same way that mild amounts of stress can cause us to up our performance level. The literal meaning of the word *tapas* is "to heat" or "to cook." Tapas is a positive stress that helps create resilience and strength in us, both inner and outer. Tapas should create energy, not exhaustion, so if you find that your practices are depleting you, then you are trying too hard, and they will cause chronic inflammation rather than growth. Swami Hariharananda says in his commentary on the *Yoga Sutras* that tapas is desisting from actions that "bring momentary pleasures, and putting up with the resulting hardship." So while tapas has an element of physical activity, it also has an accompanying practice of the things that we resist from doing as well. The physical actions of tapas should lead toward mental equilibrium, or stability of mind.[9] The practice of resisting momentary pleasures increases our willpower, and also gives us the space to question, Do I really need this thing or experience right now? Will it bring lasting happiness, or is it just momentary and therefore will end in disappointment? Through practicing austerities, the veil of impurity is removed, which means that the things clouding our minds or vision are cleared away, we become more clear, and we can see which way we need to go.

Svadhyaya is repetition of mantras or study of sacred texts. *Sva* means "one's own," and *adhyaya* means "chapter." Svadhyaya therefore is the study of the chapters on the self, the true self, or self-knowledge. The chapters on self-knowledge refer to the Vedas and the Upanishads, the Hindu texts on liberation, dharma, and philosophy. Each Brahmin family has a portion of the Vedas associated with their lineage, and the recitation of those mantras each day is called svadhyaya. An etymological derivation is *sva-adi-ayana*, which means "walking a road toward one's self." The chanting of the syllable *om* is also the practice of svadhyaya. A modern translation of *svadhyaya* is "self-study or examination," but this translation is not found in the yoga texts. In the *Yoga Sutras*, the verse says that through the repetition of mantras and study of texts, our minds merge with the deity, or level of consciousness, that we are devoted to.[10]

Ishvara pranidhana is surrender or devotion to God, or, if you are not a theistic person, surrender to the unknown. Surrender is understood as surrender of the results of all of our actions to God, or whatever view of the unknown you hold. The reason that we offer the results of all of our actions to an idea of God or the unknown is that we truly do not know what the result of every action of ours will be. While we may have a pretty good idea what some of the results will be, and have a very good idea about what others may be, we do not know the entire web of how our actions affect not only us but those around us. Actions and the results of actions are not limited to us, as independent individuals. Our actions affect other people and the environment and situations we live in. It is better to try to do the right thing, and offer the results of our actions with a feeling of unknowing toward what will come. If you are theistic, you can keep the feeling within you that God is the doer of all things, and we are His or Her instruments, and that you are bringing yourself into alignment with the highest good. The idea of surrender is not that everything we do is the will of God, but that everything we do is worthy of God.

If you do not believe in God, then practice surrender simply by offering the results of your actions to the unknown, or to the universe, with the idea that we truly do not know what reality is, what nature is, or where the universe came from. There is so much we do not know, including what dark matter and energy—which makes up 96 percent of our universe—are. We live in less than 1 percent of the known universe, and even that tiny percentage is unimaginably vast, and a great mystery. We know so very little about it.

In the *Yoga Sutras*, it is said that devotion to God is the one niyama that leads directly to samadhi, or complete absorption in pure consciousness.[11] But the sutras also say that it is only possible to attain that when all of the other yamas and niyamas have been faithfully followed!

The yamas and niyamas are ten in number and are sometimes compared to the Ten Commandments. The Ten Commandments, however, are the word and law of God, given by Him to Moses as His word to be followed. On Judgment Day, those who have faithfully followed the commandments will be allowed into the Kingdom of Heaven, while those who have not, who have sinned, will be cast out.

In yoga, there are no God-given moral commandments; the yamas and niyamas are instructions from the yogis and rishis who discovered their efficacy as they practiced and had deeper experiences of themselves in their meditations. As the original yogis gained higher and higher levels of enlightenment, the natural expression of their soul took the form of the yamas and niyamas in their behavior, and that is what they passed on. The yamas and the commandments, however, are both moral in character, meaning they encourage us to express our innate goodness in all of our actions in the world; to be decent and principled, to act with honor, kindness, and integrity. In short, everything that represents the most elevated characteristics of human behavior. They show how we can behave in this world with thoughtfulness, care, and authenticity. They are also points of reflection for us to check and see how honest we are being with ourselves about how we live in the world. Yoga, in essence, is a way for us to check how we are living. The commercial yoga industry has tried—successfully—to sell us yoga as a lifestyle, complete with products to complement it. Rather than a *lifestyle*—which is no more than another trapping, another outfit, another projected image—we want the *style or way of our life* to be more conscious. The yamas and niyamas help by giving us points of conscious behavior to meditate on. We want a lifestyle check, not a new false identity.

There is one other area where the commandments and the yamas and niyamas differ. In yoga there is the concept of karma, or action. How we behave will determine the outcomes of our life. The final judgment is not God's, but our actions will be their own judge, so that in the end it is our behavior that is the judgment and consequence. While God, or Ishvara, does appear in the *Yoga Sutras*, it is a God that is optional. You can have belief in it or not; the important thing in yoga is that you use whatever tools you need to in order to help calm and steady your mind. If belief in God helps you to steady your mind, then you should apply that belief in your life and practice. If not, you do not need to worry about it. This is another area that yoga excels in: there is room for everyone, regardless of your spiritual, religious, agnostic, atheistic, pantheistic, polytheistic, henotheistic, or monotheistic beliefs. If your belief helps you to calm and control your mind, experience your innate inner goodness, and love everyone, then you are on the right path. If your belief sets you apart from other people and leads to fights, anger, violence, judgment, and self-righteousness, then it's not yoga.

As mentioned, tapas, svadhyaya, and Ishvara pranidhana taken all together are called kriya yoga, the actions that lead toward preparing one for the attainment of the state of yoga—equilibrium of mind. They are qualified as actions that *indirectly* lead to this state of mind, meaning that we are not really working to route out the seeds of thought themselves, but with auxiliary ideas, such as doing postures or breathing, chanting mantras, or offering all of our projected ideas and desires mentally to God, all in order to calm, cleanse, and clear the mind so that we move our awareness inward. Tapas is related to physical actions, svadhyaya to verbal, and Ishvara pranidhana to mental ones. These are the primary modes that we undertake in the beginning stages of practice. The later stages of meditation work directly on uprooting the source of thoughts, but the preliminary practices of kriya yoga begin to show us that when we expand our awareness to deliberate actions, words, and thoughts, we can gain some insight to how we can be responsive rather than reactive, and use our actions, words. and thoughts as ways of being reflective and observant.

To summarize:

* The yamas are the responsibility and choices we make toward our interpersonal interactions, which are guidelines of kindness, honesty, non-stealing, sexual responsibility, and not-grasping.

* The niyamas are the choices we make for physical and mental discipline, which include cleanliness, contentment, and kriya yoga: practice, chanting and study, and devotion.

* Devotion to God is optional but helpful.

* The actions in yoga that lead to an increase in knowing who we are, and a decrease in the mental obstructions, are practice, study, and devotion.

eight

{ INTERNAL ENERGY }

AS WE MOVE DEEPER in a search for inner awareness and peace, we'll go through layers of subtlety. Our psyche is complex and contains within it the impressions of every joy, success, failure, love, trauma, and struggle that we have ever experienced. The physical feelings associated with these experiences are also stored in our nervous system and in our muscle tissue. Many people experience the release of emotions or the recollection of old memories from practicing yoga postures. The reason for this is quite simple: our body remembers, sometimes much more vividly than our mind. Trauma and also the consistently low levels of stress that many of us experience cut us off from our bodily sensations. Bessel van der Kolk, in his brilliant book *The Body Keeps the Score*, says:

> Only by getting in touch with your body, by connecting viscerally with your self, can you regain a sense of who you are, your priorities, and values. Alexithymia, dissociation, and shutdown all involve the brain structures that enable us to focus, know what we feel, and take action to protect ourselves. When these essential structures are subjected to inescapable shock, the result may be confusion and agitation, or it may be emotional detachment, often accompanied by out-of-body experiences—the feeling you're watching yourself from far away. In other words trauma makes people feel like either *some body else*, or like *no body*. In order to overcome trauma, you need to help get back in touch with *your body*, with *your Self.*[1]

While everyone has experienced some level of trauma, not everyone is traumatized. Some people have an inner resilience that allows them to either move past or grow from traumatic experiences. However, any level of trauma is held in the nervous system, and creates some level of dysfunction in pretty much everyone. By getting back in touch with our bodies, we begin a journey toward getting back in touch with ourselves through reregulating the functions of the nervous system, which are the very mechanisms that allow for a strong sense of self: "This is who I am. This is what I think. This is how I feel. This is what I want; this is what I need." Interestingly, the Greek word *psyche*, which we often use to indicate our mind, or the emotional state of our mind, actually means "soul" or "spirit" or, most tellingly, "the breath of life."

Practicing yoga postures is a very effective way to reestablish a connection with the body we inhabit, but the physiological mechanism that deepens the process of restoring the nervous system to balance is the breath. There is an integrated technique within breathing called *bandha*, which we will cover now, that can enhance the power of the breath. Thus far we have covered the practical techniques of vinyasa, the static positions of the asanas, and components of asana called tristhana. Within tristhana are asana, breathing, and dristhi. Within breathing there is also bandha. *Bandha* means "lock" or "bind." The two most important locks we use when practicing the postures are called *mula bandha*, which means "deep-rooted lock," and *uddiyana bandha*, which means "flying-up lock." A third lock used in pranayama and also in certain asanas is called *jalandhara bandha*. *Jala* means "a mesh or net," and *dhara* means "to hold."

We spoke in chapters 4 and 5 about asanas creating lightness in the body. Using bandhas and breathing within an asana helps to create that lightness. Lightness does not necessarily mean having a lean, skinny body. Some people have very lean bodies, but when I go to lift their leg or help them in a pose, they are in fact heavy; others who have bulky bodies can feel very light and supple. The internal actions of yoga create that effect, unlike, say, body building, where the outer shape is the goal. In fact, if someone has highly developed muscles from weight lifting or exercise, his or her body will quite often be stiff and heavy and, because of that, have less internal strength. Pattabhi Jois used to say that yoga is internal exercise, which means we develop internal strength, internal purification, and internal beauty—such as a good mind, kindness, and compassion. While a by-product of yoga practice can also be a fit body, it is not the end goal; a fit body is mainly useful for ensuring good quality of life. It doesn't ensure happiness, self-esteem, or insight into the nature of self.

Though bandhas are often described as muscular contractions, they should functionally be understood as an integral and subtle aspect of breathing. Engaging certain muscles of the pelvic floor and abdominal cavities supports long, smooth, and controlled inhalations, and steady, smooth exhalations, because when the breath is calm and controlled, so is the mind. Bandhas support the breathing process, as well as creating internal strength and lightness.

Besides helping to create lightness in the body, bandhas serve several other purposes. Rajas and tamas are concentrated in the body between the solar plexus and the organs of reproduction and elimination. When out of balance, rajas creates agitation, and tamas causes heaviness. For example, when your digestion is sluggish you'll feel heaviness or discomfort in the belly, an imbalance of tamas; when you are sexually aroused, even just in your thoughts, you'll sense that in the genital region, which is an activation of rajas. Because the bandhas are done in conjunction with breathing, they also help to ground the element of air. When we are anxious or overthinking or worrying, the imbalance is of *vata*, or wind or air. The mind moves like the air—it's everywhere, but you don't always know where it is going to go. And it is unpredictable, like the wind—you can have a general idea of what the temperature and weather are going to be like outside, but those change rapidly sometimes, and you can't always say for certain which way the wind is going to blow, or what the temperature is going to be from one moment to the next. This is why the steady modulation of breathing, which is wind, is so important in yoga—it helps to steady the wind, and temperature, of the mind.

When the organs of elimination become sluggish through tamas, and digestion and elimination are not working properly, illnesses or diseases of the organs of digestion and elimination can occur, such as irritable bowel syndrome, constipation, or bladder dysfunctions. When rajas is predominant, we get distracted, are quick to anger, can be sexually preoccupied, and are easily given to impulsiveness. Balancing rajas and tamas, and moving them toward sattva, harmony, is one of the goals of yoga, and bandhas are internal mechanisms that help support this.

The first bandha we'll talk about is mula bandha. *Mula* means "source," "origin," "basis," "base," or "original," and *mula bandha* means "deep-rooted." Mula bandha is performed by lifting the muscles of the internal anal sphincter, as if you were resisting a bowel movement, or performing Kegel exercises, which are done to strengthen the pelvic floor muscles to prepare for childbirth or to address bladder problems. Mula bandha helps keep the spine straight, helps protect the lower spine while you perform asanas, and helps create lightness in the body by reducing the tamas that collects in the lower abdomen and anus. It also gives strength to the lower abdomen by strengthening the very deep, lower pelvic floor muscles. The conscious toning of the anal sphincter muscles can also be helpful for maintaining healthy bowel function, because, like all muscles, they lose strength and elasticity with age, and we need those muscles for healthy elimination. You may hear it said sometimes that mula bandha is the lifting of the perineal floor. While the perineum may engage during mula bandha, it is actually, and specifically, the lifting of the internal anal sphincter muscles, for both men and women. The easiest place to find it is at the end of a slow and conscious inhale, though some people find it easier to do at the very end of the exhale.

It is very difficult to hold mula bandha for extended periods of time because the internal anal sphincter muscles are involuntary muscles.[2] They are in a constant state of contraction, controlled by the sympathetic nervous system, which rules, among other things, activation, activity, and muscular engagement. The parasympathetic nervous system rules relaxation, growth and restoration, as well as digestion and elimination. The anal sphincter muscles relax at the time we defecate due to the parasympathetic nervous system, which send signals to relax that originate in the hypothalamus.[3] The internal anal sphincter is controlled by the brain stem and the autonomic nervous system, which is the part of our nervous system that directs functions automatically—such as heartbeat and respiration—so that we can be alive. We will cover this in depth in later chapters. For now, it's enough to know that autonomic functions can be controlled for short periods of time, just like we can slow the heart for short periods of time, and control the breath for short periods of time. When you are sleeping, for example, you cannot control the speed of your breath; the nervous system will do that for you. Anytime we control or alter an autonomic nervous system function, we are in essence hacking our nervous system and placing an intentional new functionality into it. This changes the baseline operating system of our nervous system and makes it so we are not completely living on autopilot, but have a say in the mechanism and functions of how our bodies work. This changes how information is processed and absorbed by our physical and subtle bodies, and gives us a deeper insight into why we are here. When we live on autopilot, we are not questioning why we are here, or what our purpose is.

HOW TO PRACTICE MULA AND UDDIYANA BANDHAS

As an exercise, you can sit straight and inhale a smooth, comfortable breath. At the end of the inhale, pause and lift the anal sphincter muscle, as though resisting defecation. Keep this lift as you begin to exhale, as a way of supporting the smooth flow of the out-breath; this will also help to keep your spine straight. You may find that by the end of the exhale, mula bandha has loosened a little; this is normal. In the beginning, it does not need to be held all the way through the exhale, just at the start. Then after the inhale, you can catch it again. After you become accustomed to the lift of the anal sphincter while breathing, you can add it into the practice of asanas, and you will feel how it supports the different body positions and shapes that we put ourselves into while doing yoga.

You may notice that at the end of your exhalation, when practicing mula bandha, your lower abdominal muscles begin to automatically sink inward. This is uddiyana bandha, the light contraction of the lower abdominal muscles, two inches below and to either side of the navel. The area of the navel and above the navel should not become tight. Uddiyana bandha is a little tricky to isolate in the beginning if you try to find it without breathing. One way to locate uddiyana bandha is to find the bony projections of the front of your hip bones, and then move your fingers toward the centerline of your body by about two inches. As you exhale, you might feel an area where your fingers naturally sink in as the lower belly slightly contracts inwardly. This is where uddiyana bandha is. If you inhale and allow the belly to fill up, and then exhale slowly and press down where your fingers are, you'll feel uddiyana bandha. This bandha gives lightness and strength to the body. It is also said to increase the digestive fire. Both of the bandhas are functions of the breath—they are not separate from the breathing.

Mula and uddiyana bandhas can be done in almost every asana, with some being easier than others. Downward-facing dog, for example, is an easier position for uddiyana bandha to happen in than a backbend. Paschimattanasana, the forward bend, and baddhakonasana, sometimes called butterfly, are important positions for mula bandha. The inverted poses, specifically, are where the bandhas are very important for purification of the waist, the lower belly, and the organs of elimination.

Try to practice these bandhas as often as you can remember

to while doing asanas. They are subtle, and even after many years of practice, they can come and go. Don't become obsessive about them, but bring them into your practice when you can remember. They will over time contribute toward lightness in your body and steadiness in your mind. You don't need to walk around all day long trying to do mula bandha; it will aggravate the sympathetic nervous system and cause tension in your mind and body if done separately from conscious and steady breathing patterns.

Jalandhara bandha, which means "holding of the web or mesh," is performed when the spine is straight and the chin reaches forward, then down. The lock is located in between the collar bones, just above the sternum. Although I cannot be certain, my guess is that the "web" that is being held may refer to the large bundle of vagal nerves that pass through that area of the throat. This bandha tones the vagus and is said by the yogis to stimulate or purify the throat chakra, which is related to interpersonal expression. The vagus nerve in this area also affects the baroreceptors—which regulate blood pressure— and wrap around the carotid artery.[4] Located in this same area are the peripheral chemoreceptors, which are involved in the control of respiration and monitor oxygen supply to the brain.[5] This bandha may have been used by yogis for that reason as well, because it is usually employed while holding the breath. While we are holding our breath, our body will begin to send distress signals to the brain when it is time to breathe. It is possible that when we employ jalandhara bandha during breath retention, the signals to the brain are delayed, and suspension of breath can occur for extended periods of time. Furthermore, the vagus nerve is connected to facial expressions, particularly to the corner of the eyes through the oculomotor nerves, and through the larynx to expression of emotion through tone of voice.[6] The suggestion by the yogis that the throat chakra purifies self-expression could well be valid, as the stimulation of this region does indeed have a neurological basis in its connection to emotional expression. Another function of the throat, or vishuddhi chakra, is to purify the incoming air as it reaches the lungs, transforming air into breath.

Melodic, pleasant vocalization, such as humming, chanting the sound of *om*, or the ujjayi and brahmari pranayamas help to tone the vagus nerve. At least one study, done at the Nepal Medical College in Kathmandu in 2010 has shown that five minutes of practicing brahmari pranayama is effective in lowering heart rate, systolic blood pressure, and most significantly, diastolic blood pressure, which is the measure of pressure in the arteries as the heart rests between beats.[7] This is when the heart fills with blood and gets oxygen. When our blood pressure is too high, the absorption of oxygen diminishes, thus robbing our other cells of essential nutrients. Excessive arterial pressure can lead to a narrowing of the arterial walls through a buildup of fat and plaque, sometimes even interrupting blood flow to the heart. This starves the heart of oxygen, which leads to muscle death, or what we call a heart attack.

Any of the practices that encourage healthy circulation, oxygenation of blood and cells, and a healthy heartbeat is going to help extend and maximize our lives. Yoga practices collectively can do this. It's not just one practice that leads to health, but several practices combined: postures, breathing, bandhas, and

positive emotions—like gratitude, appreciation, and humility—
all lead to overall health, well-being, and longevity.

MULA BANDHA AND THE MIND

In a yogic text called the *Aparokshanubhuti*, attributed to Adi
Shankaracharya, verse 114 says:

> *yanmulam sarvabhutanam yanmulam chittabandhanam |*
> *mulabandha sada seyvo yogyasau rajayoginam ||*

This roughly translates as

> That which is the root (mula) of all existence [namely,
> pure consciousness] is the root of the restraint of the
> mind [the field of thought] and is called mula bandha;
> it should always be adapted by the Raja Yogis,
> since they are fit for its practice.

So what does this mean? *Mula* means "source" or "origin,"
and *mula bandha*, which means "deep-rooted," refers to more
than a physical muscular contraction. The most deep-rooted
part of our psyche is our impulse toward survival. The survival
instinct is wired into our nervous system, and it operates twenty-
four hours a day, every day of our life, until our life span nears its
end, when our subtle link to the body begins to fade, and survival
functions begin to weaken their hold on maintaining bodily
functions. Survival functions include heartbeat, respiration,
blood pressure, digestion, sleep, and sexual reproduction. These

functions are controlled by the autonomic nervous system, which is housed largely in the brain stem. The brain stem is the root of the brain, the gateway of the central nervous system. It sits between the body and the brain, transmitting messages back and forth. Many of the practices that yogis do, including postures, breathing practices, fasting, continence, and controlling sleep, are aimed at transcending the survival functions. The term *survival functions* is descriptive: if we stop breathing for even a few minutes, or if our heart stops beating, if we cannot digest our food, we will soon cease to live. These functions keep us alive. So why would the yogis want to mess with that? It is because intimately tied to these functions is *asmita*, or the stories we tell about ourselves. Our physiology holds on to "I-ness" as much as our mind does. The idea of transcendence, however, means that we step outside ourselves in order to know who we are on a deeper level. It is not escapism; rather, it is the opposite. If these automatic processes keep me alive, they also are keeping me tied to a narrative. But who am I beyond that narrative? What happens if I restrain my breathing for a little while each day? What happens if I can consciously and willfully slow my heart, resist the desire or need for food or sex—who will I be then? Just as we looked at the yamas as restrictions that create a healthy boundary that acts as a groundwork for internal freedom, the practices that are aimed at the brain stem functions are also the practice of restrictions.

Survival functions are the root of many of our worries and of all of our attachments. As we discussed in the preceding chapter, our attachments stem from a false sense of self, the stories we tell about ourselves instead of searching for true

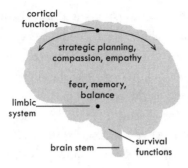

cortical
functions

strategic planning,
compassion, empathy

fear, memory,
balance

limbic
system

brain stem

survival
functions

Brain stem survival
functions include
respiration, heartbeat,
blood pressure, digestion,
sexual reproduction,
and sleep.

knowledge of who we are. Worries, fear, and attachment generate many of our thought patterns. Our mind is filled with them during the day, and at night they fill our dreams. Then there are our negative, repetitive thought patterns, the scenarios our mind concocts: imaginary arguments, imaginary disaster scenarios, imaginary romances. These patterns are rooted in the survival functions, because the thing we are trying to hold on to, the thing that we do not want to have die, is our false, limited sense of self. That is our greatest bind, our greatest identification, and our greatest attachment. Who would we be if we were not our idea of who we are? What would be left?

Yoga says that what is left over is pure consciousness, which has no location but is ever-present and all-pervasive existence. In identifying with our personal narratives, we are not ever-present or all-pervasive. We exist only where we place our mind. The mind is just a field of thought. It holds images, feelings, ideas, and fantasies, but it doesn't hold awareness, because awareness is the light that allows something to be experienced in the first place. The mind becomes a problem when it is all that we know. When we believe everything that happens in our mind is true or real, we have a big problem. When we use the mind for what it is intended for, it ceases to be a problem and is simply a tool. The mind is useful for communication, for forming words, for directing action, for turning ideas forward into something concrete. My mind is useful for forming my ideas into words, and then directing my fingers to type words on a computer; the computer is a physical manifestation of someone else's mind and imagination (mine, for example, came from Steve Jobs's imagination).

What the mind is not useful for is "figuring it all out," because "what it all is" isn't an idea, it is an experience; the mind can reflect that experience, but it cannot create it. When the mind creates an experience, it is limited; when it reflects an experience, it is an experiencer of something larger than its own capacities. This is why the mind is described as sattva—not because it is pure and harmonious, but because it has, at its root, the capacity of reflection, both self-reflection and the reflection of something else, as the moon reflects on the surface of the ocean. The sun shines light on the moon, the moon reflects that light to the ocean's surface, and we see that light and are moved by it. We respond to that light in an aesthetic way: it's beautiful, it's calming, it's poetic, it's luminous, and so forth. When we see the moonlight on the ocean, we don't think, Oh, that's the light of the sun coming from a very long distance away, at a super-high speed, crossing millions of miles of space, and bouncing its particles of light off a ball of asteroids and stardust, then landing in a random pattern on a body of water on our small planet, and that light pattern will oddly follow us as we walk down the beach in whatever direction we walk in.

Instead we think, Oh, how beautiful the moonlight is on the water tonight!

Similar to the boiling-water analogy and speed we saw earlier, this analogy points to the idea that what we experience through the senses is not all there is to it; it's not quite fully reality, just our perception of reality. The idea within the contemplative traditions is to understand the awareness that the experience takes place in, and not take experience or the sensations associated with it as reality. Within our bodies, we can see that the physiological processes are like reflections of the moon on the ocean. Consciousness reflects itself through the intellect, the mind, the nervous system, and our entire physiology. Our biological makeup is, according to the yogis, where consciousness manifests itself for the sake of having a vehicle for self-knowledge. The survival functions, whose physical location is in the brain stem, can be controlled through the conscious engagement of yogic practices and application of awareness.

The place to begin controlling the mind is in the brain stem via our survival (or autonomic) functions, as the root of the mind is housed there, in what is evolutionarily the oldest part of our brain. The brain stem keeps us alive through our respiration, heartbeat, and all of the other autonomic survival functions, and our identification as living beings is completely intertwined and identified with these functions on an unconscious level. The unconcious survival functions are one place where we can find the root of thoughts, specifically the thought "I am this body." All of the other thoughts that we have, all thoughts and ideas concerning ourselves and our existence, follow from this root thought. Shankaracharya says, in the earlier quote, in what is truly a bottom-up approach to controlling the mind (no pun intended), that the true mula bandha is the restraint of the root of the mind, the citta bandha, and not the restraint of the sphincter. He said this because the intense austerities that the yogis sometimes followed, including mula bandha, complicated postures, and long breath holds, do not always lead to liberation, but to a glorification of the body and the ability to perform extreme feats with it. Before he sat under the bodhi tree and attained enlightenment, the Buddha practiced austerities under the guidance of a yoga guru, but he eventually stopped doing them because he realized he was torturing his body without making spiritual progress. To restrain the root of the mind is to restrain the physiological correlate for where our narrative initiates, which is in the brain stem. Higher levels of thought occur in the limbic system and prefrontal cortex, but without the self-identity on a survival level that occurs in the brain stem, higher-level functions will not occur.

MULA BANDHA AND THE
AUTONOMIC NERVOUS SYSTEM

Although Shankaracharya said that the true mula bandha is restraining the root of the mind, not contracting the anal sphincter, there is an anatomically direct connection between the anal sphincter and the brain stem. Every part of our bodies is in a constant communication with every other part. The nerves that terminate in the anal sphincter are also related to our heart rate, through the nerve endings of the vagus that connect via a chain of neural connections extending from the intestines down to the anus. The vagus nerve, which we will cover more in chapter 11, is responsible for a myriad of functions, one of which is to control the slowing of the heart rate, one of our primary survival functions. If not for the vagus nerve, the heart rate would be a steady ninety to one hundred beats per minute; the braking mechanism of the vagus (the vagal brake) slows the heart down to about sixty-five to seventy beats per minute and allows for fluctuations in heart rate depending on our level of activity. This fluctuation, called heart rate variability, is a measure of the healthy functioning of our nervous system. There are certain times, however, when stimulation of the vagus can be dangerous. Gastroenterologists, for example, know that they have to be careful when they dilate the anus in small children because it can lead to bradycardia, a dangerous slowing of the heart rate; the same is true when intubating a baby or small child for anesthesia. When the epiglottis is moved to place the tube down the throat, the heart can also go into bradycardia.

It's interesting that the anal sphincter and the epiglottis both can contribute to the slowing of the heart. By controlling both the anal sphincter through mula bandha and the epiglottis through jalandhara bandha, the yogis found that it was possible consciously to slow the heart rate, and that when the heart rate slows, there is a correlative emotional and mental effect. When the heart rate is consciously slowed, the vagal brake is on and the mind becomes extremely calm, focused, and inward. That enables us to select and examine the thought patterns we wish to have in the field of our mind—even the thought pattern of having no thoughts. There is a direct link between physical practices and mental and emotional effects. It seems that the yogis figured out the correlatives in these physiological principles, and applied them toward spiritual evolution. Their bodies were not taken for granted, but were a jumping-off point for spiritual exploration and understanding, for investigation into reality, nature, and consciousness.

The physical practices of yoga—asanas, pranayama, bandhas, and the like—enable us to investigate where consciousness and our personal biology meet. How do I find or discover where consciousness is in me? We start off thinking that consciousness has a location, and in the beginning it indeed does. It is tied to our bodies. When we feel or identify with primarily being our bodies, then our lives become centered around our bodies as being who we are, and all other bodies as being separate. When we view the bodies of other people as being "other," then we objectify them as we objectify ourselves. Other bodies can then become a source for competition, pleasure, or something to conquer. Our mind and nervous system, and perhaps our sense of purpose, are subsumed by this feeling of separation, and then serve the pleasures or desires we associate with our body (such as food, sex, and shopping, which are not bad things, but perhaps not a purpose enough for living). When we only live outwardly, we don't develop an inner life. We usually begin to feel the pull toward yoga or meditation practices when we feel that we need to begin developing an inner life. As we practice asanas, the awareness we have of our body expands; we can experience our body in new ways, and feel where our thoughts, emotions, and ideas are held. And then, as the body and nervous system open, we can release things that have held us tightly. As this happens, our consciousness expands, because expansion naturally follows release. Automatically compassion, empathy, understanding, and forgiveness express themselves in us, because they are the traits of a mind that is not self-centered around an individual body; they allow us to become more connected to the world and people around us. This is what is meant by consciousness as a non-local experience, when our circle expands beyond our own needs, beyond our own body. We can move from the experience of consciousness as local and fixed to the body and mind, to one of consciousness as expanded or infinite connection. Yoga states that there is a biological basis for this, and the purpose of our bodies is to have that experience. In yoga, the body exists so that we have a vehicle to use to fulfill our purpose in life, whatever that may be. Therefore, all bodies are sacred, because they exist to fulfill each individual's reason for being.

n i n e

||||||

{ BREATH AS SPIRIT }

THE LINK BETWEEN THE BREATH and the spirit is at the root of nearly every single contemplative and religious practice. Verses about the breath as spirit occur in the Bible:

> Then the Lord God formed a man from the dust of the ground. He breathed the breath of life into the man's nostrils, and the man became a living person.
> —GENESIS 2:7

The God who made the world and everything in it is the Lord of heaven and earth and does not live in temples made by human hands. Nor is He served by human hands, as if He needed anything, because He Himself gives all men life and breath and everything else.

—ACTS 17:24–25

My words come from an upright heart; my lips sincerely speak what I know.
The Spirit of God has made me; the breath of the Almighty gives me life.

—JOB 33:3–4

In Sanskrit, the word for "breath" is *prana*. It is given a status of utmost importance in the yogic texts and philosophical treatises, including the Upanishads and the *Bhagavad Gita*. *Prana* has many meanings. Its literal Sanskrit derivation is *pr*, "that which comes before," and *ana*, which means "breath." That which comes before the breath is the impulse toward life. We do not know where this impulse comes from, or how it manifests within us. Breath is indeed one of the most mysterious facets of our existence. Though the act of breathing can be explained by atmospheric pressure, how is it that breath causes us to live and, when it leaves, causes us to die?

In returning to the *Chandogya Upanishad*, there is a narrative that describes how prana is supreme among all the powers that we have as humans, greater than our mind, greater than the elements, and greater even than hope or reflection. It takes the form of a conversation between the sage Narada, who asks,

"What is the greatest thing in the universe? I have studied so many things, but they have not revealed to me the ultimate truth," and the guru Sanatkumar, who answers that what Narada has learned are names only. Sanatkumar explains:

> *Greater than name is speech, because speech leads to understanding;*
> *Greater than speech is the mind, because the mind is infused with the self;*
> *Greater than the mind is will, because will has caused all things to be;*
> *Greater than the will is intelligence, because in intelligence name, speech, mind, and will have their basis;*
> *Greater than intelligence is contemplation, because contemplation is tranquility;*
> *Greater than contemplation is understanding, because understanding is necessary for contemplation;*
> *Greater than understanding is strength, because through strength all things stand steady;*
> *Greater than strength is food, because without food one cannot be steady, see, hear, reflect or understand;*
> *Greater than food is water, because without water, food and all things cannot grow;*
> *Greater than water is fire, because fire shows itself before water in the form of heat and lightning;*
> *Greater than fire is space, because all things exist only in space;*
> *Greater than space is memory, because through memory perception is possible;*

*Greater than memory is aspiration; because memory is
    kindled by aspiration;*
*Greater than aspiration is prana, because prana causes the
    spirit to reside in the body."*

There is nothing greater than prana. As the spokes of a wheel are all attached to the hub, so are all things attached to prana. Prana gives vitality to the air, it gives animation to all living things. Sanatkumar continues by saying:

> *"Prana is the father, prana is the mother, prana is the
> brother, prana is the sister, prana is preceptor, prana is
> the Brahmana (the knower of truth)."*[1]

The animating force of prana is the basis for everything else on the list, and one who worships prana, and who comes to know prana, knows everything else that there is to know.

Later, in this same Upanishad,[2] there is a story about the sense organs arguing over who is supreme among them. Each of the organs, including the mind, leave the body for a year to challenge the others, but even absent a sense organ, life goes on, and they adapt. Finally, all of the sense organs start to feel their power waning, as if they were evaporating, and realize that prana is withdrawing itself. Prana says, "It is by my power that any of you exist, and when I leave, you dissolve. I am the underlying power of breath, energy, attention, awareness, and the power that invigorates the sense organs." The sense organs all bow to prana and say, "Indeed, you are supreme among us, please do not leave or we will perish!" Prana is the underlying mechanism of coherence, it is the glue that holds us together and binds our organism and identity to a time, place, and form. Prana binds us to the illusory nature of personality and narrative, but prana can also unbind us from all transient identifications, by leading us toward inner identification with awareness.

In the *Taittiriya Upanishad*, the first written description of the five bodies (which were discussed earlier) is given. Each body contains within it five parts. The body is described as a bird, with a head, trunk, two wings, and a tail, all of which fit and work together as a uniform whole. The word *maya* is used to describe each of these bodies. *Maya* means "mist" or "cloud." The bodies are not separate, discrete entities, but componants of a whole. On my first trip to Greece with my wife and daughter many years ago, I was struck by the image of a downward-flying dove at the entranceway to many of the Greek Orthodox churches. The same image the Eastern Orthodoxy uses for the holy spirit, or breath, is the same image used in the Upanishads for the inseparable nature of our five-fold bodies and the prana that links them. The descent of the holy spirit, or breath, into the body binds us to our earthly existence. In this model, the five-fold body is a vehicle of consciousness. The function of yoga is described at the level of buddhi, or intellect, where the Upanishad says:

> *Conviction (shraddha) is his head;*
> *Cosmic righteousness (rtam) is his right wing:*
> *Truth (satyam) is his left wing;*
> *Yoga is his trunk;*

Cosmic intelligence, the Universal sense of "I" (mahat), is his support.[3]

Yoga here is the binding force. It links together universal "I-ness" with the highest expression of individual consciousness as truth, faith, conviction, and alignment with *rtam*, or the harmonious balance of nature, with cosmic intelligence as its base and support. Yoga in these older Upanishads is poetic and evocative much more so than explicit and instructional; however, what is directly stated here is that yoga harnesses the power of the intellect and links it to its support, the *anandamaya*, or body of bliss. What is the body of bliss, the support of the intellect, the body that lies closest to pure being?

Love (priyam) is his head;
Joy (modah) is his right wing;
Delight (pramodah) is his left wing;
Bliss (ananda) is his trunk;
Consciousness (brahma) is his foundation.

Consciousness is described here as the foundation of everything, including all of the other bodies. The veil between pure consciousness and the body that lies after it, the intellect body, is thin, and love is the very thing that helps to pierce the veil of the intellect. The Upanishads state that we cannot experience the inner self through force, or through great feats of strength, or through complicated yoga poses or intellectual achievements; it is through love that we pierce through the veil that covers the bliss of being and connectivity.

Indeed, prana and awareness are linked together. Prana flows where we direct our awareness, and vice versa. Practices such as pranayama, in particular, bind prana and awareness together, or breath and awareness together, so that as the breath becomes very thin, subtle, and slow, the mind becomes very quiet as well—meaning the contents of the mind, which are our thoughts, memories, and the like. According to yoga, thinking occurs when the breath moves independent of awareness, but when breath and awareness are harmonized, discursive thinking can be controlled. Breath is considered to be of utmost importance, as it can lead to control of the mind. In the *Hatha Pradipika*, there is a famous verse that says:

> *When the breath moves, the mind moves. When the breath is still, the mind is still. By thus controlling the breath, the yogi attains steadiness.*[4]

Calming and slowing of the breath is directly linked to the state of our mind and emotions. A recent study, in fact, has linked respirator phases to attentional performance.[5] We all know that when someone is upset, angry, or in a panic, we tell that person to breathe. We know intuitively that breathing will calm whatever upset we are going through, because when we are in a state of upset, our breathing changes. It becomes uneven or jagged, or we hold it. The basic project behind yoga is the stilling of the mind, and the yogis used the breath and different types of breathing practices to quiet the mind. It's been shown in studies on Tibetan monks and yogis that when they are in meditative states, their breathing patterns automatically shift and slow to a calm, even rate, somewhere between five to seven breaths per minute, down from a normal breathing rate of fifteen to eighteen breaths per minute.[6] Breathing calmly will quiet the mind, and quieting the mind will calm the breath. However, it's much easier to start with controlling the breath than to start with the mind, because our breath is easier to manipulate. We can feel it, we can sense it, we have a certain degree of physical control over how we allow it to move. But the mind—we can't see it, we don't know where it is, and it moves however it likes to, quite often against our own will and desires. The yogis suggested working with something that we do have a measure of control over, like the body and breath, and from there, going deeper.

As we explore this connection we can begin to see what the underlying mechanism of yoga very well may be, and why yoga is helpful for so many different types of problems. In the last chapter, we'll go into the neurophysiological mechanisms of the breath and nervous system, and how from a scientific point of view the breath is linked to our mind and emotions. It is a little bit complicated, but it gets to the heart of the matter: how states of consciousness are an inherent part of our physiology. Why should we not, asked the yogis, use our physiology to manipulate ourselves into the higher states of consciousness that we wish to be in? In the next chapter, we'll cover practical tips on how practice changes our physiology, and what we can do to support that in a healthy, balanced way.

t e n

||||||

TIPS ON PRACTICE

*What we practice makes us
become who we are next.*

—JOSEPH GOLDSTEIN

ONCE WE ESTABLISH A YOGA PRACTICE, we on occasion find that we are not always as regular, consistent, and diligent as we would like to be, and a couple of things may happen:

* We can get discouraged and give up.

* We can become hyperenthusiastic and then fail to keep our momentum going through cycles of practicing and not practicing, until we find one day we aren't doing it anymore.

The *Bhagavad Gita* provides a simple formula for how to maintain a consistent, diligent, and, most important, effective yoga practice.

Yuktahara viharasya yuktacestasya karamasu |
yukta svapna avabodhasya yoga bhavati duhkhaha ||

||||||

For that person who is moderate in food, moderate in enjoyments, moderate in work, and moderate in sleep, yoga is the remover of suffering.[1]

Even twenty-five hundred years ago, moderation apparently was the key to a happy life. We often think life was easier in the past, because there was no technology, no stressful city living, and so forth . . . but of course that isn't true. As long as there have been people—especially people living in cities—there has been stress. In this chapter we'll go over some basic suggestions and ideas to help you build a consistent, moderate, and effective yoga practice and develop a healthy lifestyle that will support a yoga practice, namely, how and what you eat, how well you sleep, and what you do to relax and enjoy life.

What's interesting about this verse is that though we know that we need discipline in yoga and that without it our progress is slow or does not come at all, it tells us that we also need to be moderate and enjoy life. Discipline does not mean rigidity. It means that we recognize that what we are doing is important enough to be committed to; once we recognize that, we choose our level of commitment, as in any relationship. The more we enjoy what we choose to practice, the more likely we actually are to do it. Your practice should be something you are passionate about. Even if it is challenging or difficult to do at times, it should bring you joy or fulfillment or a feeling of satisfaction that you have attended not just to your body and mind but also to that invisible part of yourself that is the essence of who you are. So while yoga is a discipline, we also have to make sure that we love doing it. That love for practicing will make our efforts at being disciplined softer and kinder. And then we will become that way, too.

IN REGARD TO YOGA PRACTICE

1. Decide how often you want to practice. Two, three, four, or five times per week—it is up to you. Even once a week is okay as long as you actually do it. Allow this to change over time. A daily practice might be too much for you when you first start doing yoga, but perhaps after some time, it will be second nature for you to wake up every day and do some practice.

2. Choose the days that you will practice, and try to stick to them so it becomes a part of your routine. If you are practicing daily, make sure you leave at least one day for rest.

3. See if you can practice at the same time each day. This is very important for forming a new internal

rhythm, and will help to hardwire your new habit in you.

4. If you struggle to make it to class on your own, bring a friend or find someone who is also interested in making practice a regular part of their life. Community, called *sangha* in Sanskrit, is helpful for maintaining regularity in practice. Our sangha becomes our spiritual friends.

5. Appreciate and give silent thanks each time you practice. Congratulate yourself each time you practice. And when you finish, reflect on your efforts and let the feeling of that soak into you. In this way you will bring your practice into your long-term memory and it will become a part of the background character trait of your conscious mind.

6. Recognize that practicing yoga is good for you. It's a time for you to be with your thoughts, your body, your breath, and your potential to expand your capacities, all necessary things for us to take the time for.

7. If you find yourself getting obsessive or compulsive about practice, back off a little. If you are not able to temper yourself in your actual practice, then you may need to take a little break, or relax your discipline slightly. Eat some chocolate, go to a movie, sleep in. As soon as we become too driven about practice, we reinforce old patterns. At the

same time, you have to watch out for laziness. Skipping one day of practice is okay, but be careful, because it can lead to skipping two days, then three, then many!

8. Yoga should create a feeling of vitality in you. Try to do your practice in such a way that you feel you are building energy, not depleting yourself.

9. It is not that difficult to build energy through yoga, but it can be difficult not to waste it. Considering lifestyle changes and examining addictive tendencies will go a long way toward preventing you from wasting your newfound energy.

In the *Bhagavad Gita*, Lord Krishna says that practice that is like poison in the beginning but filled with nectar or joy in the end is sattvic; practice that is like nectar in the beginning but like poison in the end is rajasic; and that which is like poison in the beginning and at the end is tamasic. Check to make sure your practices all fall into the first category. And of course, after some time, our practices can become a joy at the beginning *and* at the end. But sometimes they are just plain hard. That's okay, too, and that's part of why it's called practice. Because regardless of how today went, tomorrow we have to get up and try again.

Yoga and enjoyment of life do not need to be mutually exclusive. It's important to have friends, to have hobbies, and to learn new things. All of these are good emotionally and physiologically; our brain structure changes when we continue to learn new things. It grows new neurons and keeps us active and sharp even as we get older. When you embark on a spiritual path, try not to neglect the enjoyment of friends, of learning new things, of enjoying music, art, literature, and time spent in nature. All of these are important because they help us step outside our own minds and problems for a little while each day.

Enjoying life sometimes means removing yourself from the things in life that are difficult, not as a way of avoiding problems, but as a way of remembering that our problems are not the only thing there is in life. For myself, music is my main enjoyment escape. Music is good for us on many levels, from the effect it has on the brain to how it draws us into a shared emotional experience with other people. It is uniquely human. Though the wind, rain, oceans, and stars all create their own symphony, music created by humans is unique because we make it for each other. The rain patters on the earth, creating a song, and the earth absorbs that rain, heat energy later evaporating it back into the clouds. But who is it that appreciates that rain pattern? Does the soil? Do the trees? Do the clouds enjoy the sound of rain? Or is it just the human ear that takes that sound of rain, the song of rain, and turns it into music, lyrics, and emotionally uplifting movies, like "Singin' in the Rain," or songs, like Led Zeppelin's "The Rain Song," that are enjoyed by millions?

IN REGARD TO FOOD

One of the objectives of practicing asanas and pranayama is to increase the power of digestion. Strengthening our digestion improves both physical and emotional health, as the enteric nervous system, or nervous system in the gut, links the intestinal functions with the emotional and cognitive centers in the brain.[2] In yoga, the element of fire rules digestion, both of food and life experiences. When we can digest food, our body becomes nourished, and we have the energy that we need to do the things in life that we want to do. When we are unable to properly digest food, it can lead to a host of problems, including diarrhea, hemorrhoids, headaches, and low energy, to name just a few. Certain types of unhealthy diets can actually lead to digestive disorders, such as acid reflux or irritable bowel syndrome (which can also be triggered by stress, which is the inability to manage or "digest" life experiences). The microbiota of the gut, which responds very quickly to dietary changes and stress levels,[3] has actually been linked as a driver to a variety of illnesses, including vascular dysfunctions and certain cancers—more on this below.[4] Even general indigestion can be troubling for daily life. Digestion of experiences means that we can handle the things that life throws at us. When we do not have the reserves to deal with our everyday pressures, stress begins to accumulate within us, and we get "stressed out," which is another type of poor assimilation, and as we all know, stress drives us to take comfort in foods that might taste good but are not nourishing. Circling back to the meaning of yoga as relation, we can add food into the equation, along with postures, breathing, sleep, meditation, and behavior. When trying to shape a new relationship with food, a simple introduction about not only what to eat, but how and when, can be helpful. Here are a few ideas.

WHEN TO EAT

In regard to yoga, it is important to practice on an empty stomach. All of the twisting and bending forward and backward will be uncomfortable with food or drink in your stomach. Your last meal before yoga should be between three to four hours before practicing, but you can have a piece of fruit or an energy bar up to one hour before. Coffee or tea are fine thirty minutes to an hour before practice.

Exercise and digestion are two separate metabolic processes, as are sleep and digestion. It is also beneficial to have your last meal two to three hours before bed. Rest and digestion are parasympathetic nervous system functions, so you can also expect that when you have a bigger meal, you are going to get sleepy. If you have a big afternoon of work ahead of you, eat lighter, so you are more alert. If you have a big meal before bed, your nervous system is focusing on digestion, rather than sleep, so your sleep will not be as deep or restful as it might be.

WHAT TO EAT

The yoga texts divide food into the three categories of the gunas.

Foods that are sattvic promote vitality, energy, vigor, health, joy, and good appetite. They will be fresh in nature and appearance. Think fruits, nuts, vegetables, beans, and grains—lightly spiced, tasty, easy to digest, sweet, not too oily, and freshly prepared. Potato chips, for example, are not sattvic, though they are tasty. Look for foods that are high in fiber and can be freshly prepared.

Foods that fall into the rajasic category are bitter, sour, oversalted, over-spicy, pungent, dry, or burning. They are productive, according to the texts, of pain, grief, and disease. While it is nice to eat spicy food every once in a while, don't make it your mainstay. If your eyes are watering, your nose is running, and your head reels at every meal, you are not eating nourishing food. An overabundance of this kind of food is said to be a hindrance to spiritual practice, because it makes the mind overactive; it is also not great for your digestive organs, and over time can damage the lining of the stomach walls.

Tamasic foods are stale, tasteless, smelly, old, cooked and a day or more old, or rotten. All these foods should be avoided, for the most part. This includes fried foods, which are quite often prepared in oils that have been boiled many times over. Tamasic foods lead to heaviness, poor digestion, and laziness. If you walk down the street at an Italian street fair in Little Italy in New York City, you'll see a wide range of examples of tamasic food. Foods that are high in fat, carbohydrates, and processed sugar should be taken in very small doses. But it is of course better to not have them at all, because of their effects on the health of our blood and because the microbiome will respond to their influence immediately.

The microbiome consists of the bacteria that form the lining of our gut and are found on the skin, in the mouth, and in the vagina. These bacteria were earlier thought to outnumber our human genetic material by about 99 percent microbial cells to 1 percent human cells. This estimate was from a study done in

1977; however, newer calculations have shown that the amount of microbiota is roughly the same as the cells in our body—between 37 to 40 trillion, depending on who you ask.[5] However, this does not discount their importance. The microbiome is very adaptable and can change rapidly depending on the foods we eat.[6] A healthy microbiome will lead to a stronger immune system, lower levels of inflammation in the body, and stronger digestion. Microbiomes that are exposed to damaging influences such as low-fiber diets, stress, anger, and unhealthy foods high in fat and carbohydrates have been shown to be correlates to heart disease, digestive disorders, inflammatory problems, and mental illness.[7]

The microbiome is a very big topic today, and more information can be found on the microbiome in the many books that have been written in the past few years. Among them are *Super Genes* by Deepak Chopra and Rudolph E. Tanzi, a discoverer of three of the genes that cause Alzheimer's, and *10% Human* by Alanna Collen. Collen is an evolutionary biologist who went down the rabbit hole of the microbiome after being bitten by a tick in a tropical rainforest while studying bats. She discusses how the microbiome and humans have evolved together over many tens of thousands of years, and that the microbial environment in our gut affects not just our digestion but our physical and mental health as well. It is worth becoming familiar with the notion that we have about three pounds of microbes in us—and that is the same weight as our brain.

Both the gut and the spinal cord have their own independent neural networks. These two environments, the brain and the gut, the latter of which contains the microbiome, are extremely active in how we relate to, and are an integral part of, our environment, and how our environment, in turn, is shaping us. Our brain function is not just shaped by incoming inputs from the sensory organs; it is also shaped by the messages that the microbiome is sending up to it, via the vagus nerve. This is actually one of the best reasons for a healthy diet. The food we eat dictates messages sent to the brain, influencing our emotional and cognitive functions. Unhealthy food choices, or foods that may be healthy but are not ones that we can easily digest, can lead to low energy, depression, heart disease, and poor cognitive abilities. For example, bell peppers are rich in vitamin C and antioxidants, and can be a good addition to a healthy diet. But for some reason, my digestive system cannot handle them, and anytime I eat a bell pepper, my digestion is thrown off. Just because something is healthy, it doesn't mean that it will be good for every individual. In Ayurveda it is said that it is not what you eat but what you can digest that leads to vibrant health. Experiment with different foods, observe their effects on your system, and seek guidance on diet when you need it.

Fad diets may work for some, but not all. Digestion is an individual experience, and sometimes you have to play around a little with your diet. Sometimes simple adjustments to diet can lead to significant improvements in someone's overall feeling of well-being. For example, for a lot of people, cutting out one hard-to-digest food like wheat, dairy, or processed sugars can lead to instant changes in digestion, elimination, energy levels, and mood. I've seen yoga students have huge transformations in their bodies and in the level of joy they obtain from their practice by making a simple change like cutting sugar or wheat from their diet. These changes are all individualistic: what works for one may not work for another. However, for most people, small tweaks can lead to big changes. All of this falls under the jurisdiction of the gut-brain axis, and one of the amazing things about this system is that it can respond very quickly to positive changes.

HOW TO EAT

Here are the basic suggestions found everywhere:

* Eat sitting down.

* Eat at regular times.

* Eat slowly.

* Eat with friends or family when possible.

* Appreciate your food and give a silent thanks for the nourishment food gives you.

* Try not to waste food.

* Eat slowly, and enjoy the taste of your food. Enjoying the taste and appreciating the preparation that went into your food helps digestion.

Take a look at the foods you eat each day, and see if you can identify which categories they fall into. Try to eat more on the sattvic scale; a little rajas like tea, coffee, spices, chocolate, and sugar is okay, but they should be minimized. For example, six cups of coffee a day is too much. Try to avoid all tamasic foods, except perhaps on special occasions. They just plain aren't good for you.

A vegetarian diet is certainly good for your health, energy, and mental capacities—but it is not for everyone, or at least not right away. Learn to be discriminating about what you eat and when, and avoid rigid regimes. They are hard to stick to and can make us ideologically rigid as well. Of course, certain health problems may require a disciplined diet and avoidance of certain foods.

SLEEP

When we have a balanced relationship between food, play, work, and sleep, our minds are clearer and it is much easier to be in touch with our sense of self, purpose, and direction. It's important to understand that moderation does not mean we need to live like monks or in constant calibration of every move. It means that our extremes should be less extreme and that we should become more sophisticated in catching ourselves when we need to set our rudder straight, and have the tools to do so. It's important to enjoy life, but it's also important to enjoy work, sleep, food, and yoga. Now that we've looked at food in relation to the gunas, we will look at the gunas in relation to sleep. This will help give us some guidelines or suggestions to us about how best to feel rested.

Sattvic sleep is that which is deep, the right length of time, and free from disturbing dreams. According to Krishnamacharya, sleep is said to come at night when we stop thinking. Everyone knows that the more we think, the harder it is to sleep. When our sympathetic nervous system is amped up, we are mentally more active. Daily yoga practice, especially the daily practice of resonance breathing (see practice A), meditation, and deep relaxation at the end of practice, trains us to down-regulate sympathetic nervous system functions and stimulate parasympathetic nervous system functions. This is why it is called conscious rest. If you find that you lie in bed at night thinking, gently remind yourself that sleep will come if you can let your thinking taper off and use tools like a body scan (see practice D) or conscious relaxation suggestions to sense where sleep is in your body, and allow yourself to sink into that place.

The right amount of sleep is generally thought to be about seven to eight hours per night. The brain needs this amount of time to repair itself each night, as do the bodily tissues and muscles and especially the immune system. There is a special system of glial cells in the brain, called the glymphatic system, that actually clears the brain of the debris that collects throughout the day from thinking. Yes, that is right—thinking actually leaves a physical residue that can turn into a plaque-like material if not cleaned from our brains each night.[8]

Dreamless sleep, or sleep with pleasant dreams, is always an enjoyable experience, especially upon awakening. For many of us, that kind of sleep is a rare experience. One theory about dreams is that they are the experiences we have had during the day working themselves out—those that we have not assimilated (digested), expressed, or come to terms with. They replay so that our unconscious mind can make sense of them, and play them out in a dream setting. Patanjali suggested meditating on the content of your dreams to help your unconscious and conscious thought patterns assimilate into each other; this will help with creating a calm and integrated mind.[9]

Rajas, which we define as *active*, *passionate*, and *energetic*, is reflected in sleep when we are overexcited or when we have had too much caffeine, sugar, or stimulants—especially computer, TV, and phone screen time. Rajasic sleep is restless, broken, and filled with disturbing dreams, or does not come at all. (Insomnia can be either rajasic or tamasic. It is one of the most unpleasant of sleep disorders.)

To correct this, we need to apply some discipline. Turn off the computer earlier; cut back on coffee and tea; avoid anything upsetting, distressing, or stressful in the evening. Meditation, body scan, or a few relaxing postures can help your nervous system settle down and prepare you to let your thinking subside so that sleep can come. If you have had disturbed sleep for some time, it can take a while for your nervous system to adjust and your body to catch up, but if you make this a disciplined habit, you will be greatly rewarded. I have found that just a few minutes of resonance breathing each day has improved my sleep, and my ability to fall asleep much more quickly. You can't try these strategies just on the nights you can't sleep; they have to be done every day so that your nervous system can begin to adjust to a new rhythm. Learning to sleep well is a practice, too.

Like many people, I am in front of my phone and computer many hours a day. That's a lot of incoming stimulus. Resonance breathing has helped me to get in touch with the parasympathetic branch of the nervous system that rules rest. When I lie down to sleep at night, I can sense sleep much more readily, and I can give in to it. It feels like the breathing practice has helped me learn to consciously activate the parasympathetic branch as soon as I put my head on the pillow, and my nervous system knows which direction to go in. For many people, difficulty falling asleep creates anxiety, especially when you are lying in bed trying to fall asleep. Activation of the parasympathetic nervous system creates a feeling of calm, and the anxiety diminishes. It's easier to deal with the underlying mechanism of sleep, namely the parasympathetic nervous system activation, rather than trying to get rid of anxiety, which almost never works. Again, many of the sleep interventions work best when they become habits, and aren't things that you do just on the nights when you can't sleep. They aren't like taking aspirin if you have a headache. You need to train your nervous system to know when it's time to sleep, and for that, you have to follow a regular routine that prepares you each night for sleep, so your body, nervous system, and mind all know what to do when you get in bed.

Depression, overwork, trauma, and poor lifestyle choices can all lead to tamasic sleep, which is characterized by extreme fatigue that can make it difficult even to fall asleep. Most of us have had the experience of being too tired to sleep—it is a horrible feeling. Tamasic sleep that is due to depression can cause us to stay in bed for days at a time and never feel rested. On occasion, this kind of sleep is a symptom of deeper psychological or physiological issues, and sometimes a specialist needs to be seen. If the symptoms are not extreme but just a case of chronic overwork, then lifestyle adjustment and introspection can help to change course so that we are not constantly exhausted. Remember: much of our lifestyle is a choice. Even when we live in difficult circumstances there are choices we can make that can lead to positive changes. Small tweaks can lead to big changes.

For both rajasic and tamasic sleep problems, introspection is a key component of change. While things like warm baths, essential oils, and turning off the computer earlier are all helpful, the most important thing for us to do is to understand clearly what our triggers are and why we are doing things that we know are not healthy or supportive. This is truly the hardest thing to do in a spiritual practice, to take a good look at why we have the habits we do and what drives us to repeat them. A little bit of self-honesty goes a very long way, and our nature, or personality, is revealed to us through the way we practice yoga, enjoy life, work, and sleep. When one of these is out of balance, then simply giving a little attention to that one thing will reveal where we need to transform ourselves, where we still need to grow and change. The practices are mirrors for us to look into. Our nature is revealed through them. We don't always like what we see, but with a positive mind-set, we can apply a little effort and will quickly begin to see positive change. That's really all it takes. A little bit of consistency, a little bit of commitment, a little bit of honesty, and we can go a long way toward both change and growth.

eleven

ꞮꞮꞮꞮꞮꞮ

THE NERVOUS SYSTEM,
EAST AND WEST

IN THIS FINAL CHAPTER, we'll look at some of the reasons why the breath and the nervous system are so important. It is a complicated topic, but it gets to the heart of why yoga works. The nervous system is massively complex, so I'll be discussing it in broad terms and offering some general thoughts about its functions and processes in both yogic and scientific terms. There will be some discussion about the anatomy of the nervous system, but mostly I'll talk about how conscious breathing and other practices affect the processes that the nervous system oversees.

In this chapter we will cover:

* An overview of the pranic and nadi systems
* An overview of the nervous system
* The sympathetic and parasympathetic nerves of the autonomic nervous system (including the vagus nerve)
* Brain stem functions
* Where the kleshas exist in the brain
* The four practices used by the yogis to harmonize brain and nervous system functions

Breath is essential to our existence, though we rarely think about it, and usually only when we start to run out of it. In fact, it is like that with pretty much every part of our body, inside and out—we don't think about our body in any real, functional way until something goes wrong with it. We might pay a lot of attention to our hair, face, and physique, but we pay very little attention to the things that actually make us go, or that really matter in terms of quality of life, like the health of our liver or kidneys or the variability of our heart rate. Our body doesn't come with warranties or replacement plans; we can't just throw it out and get a new one and start over when something goes terribly wrong. When it comes to our health and our longevity, a little consistent maintenance goes a long way.

Our body has a tremendous capacity to correct and heal itself. The ancient yogis placed emphasis on breathing practices because the breath has tremendous influence on the nervous system, which is where our body's internal balance and autocorrect mechanism—called homeostasis—lives. The breathing practices were not done independent of postures and other practices, so we will look at how all of these practices together affect the many rhythms of the nervous system. Conscious breathing, however, is one of the easiest ways to begin to balance our autonomic, or automatic, bodily rhythms and gain the skills needed to adjust them for both daily maintenance and for when they go out of balance. The nervous system coordinates most of the activities of our cellular environment, all of our physiological processes, and our emotional responses to the world around us, so strengthening and balancing the nervous system is a key component of physical and emotional health. Homeostasis is the process of the body's adaptation to changes in the environment through the responsiveness of the nervous system, and is the body's mechanism for maintaining the stability of blood pressure, body temperature, gas exchange, and blood pH, all in concert with the demands of the external environment. It has developed over millions of years of evolution. It should be noted that the body spends a tremendous amount of metabolic energy to keep us in balance and maintain homeostasis.

Yoga considers that it is one of our basic responsibilities to care for the homes that we inhabit—our bodies. It's our job to keep them internally and externally clean, well functioning,

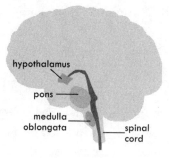

While homeostasis is regulated through a complex feedback mechanism, its primary control center is the hypothalamus, which acts on the medulla.

and resilient. In the *Dharma Shastras* it says that our first dharma, or duty, is the duty of taking care of our bodies.[1] If we do just a little maintenance every day, our bodies and everything in them will last longer, function better, and make it possible for us to take care of our families, contribute to society, and make our lives more fulfilling.

Our ability to regulate the nervous system, to achieve homeostasis, is a large part of our ability to be resilient, which means that we are able to bounce back after regular day-to-day challenges as well as from illness, fatigue, emotional stress, or trauma. All of the practices we've discussed so far in this book help increase resiliency and help us come into balance and feel energized on a daily basis. When our system is fatigued by any of the stressors listed above, the length of our recovery time is slower, and it can indicate that we may need to take a pause and give our system time to come back online. As we look deeper into the various mechanics of yoga practices, we find that developing increased regulation of the nervous system—which

occurs through practices such as asanas and pranayama—is a
key commonality.

THE FIVEFOLD PRANA

From the yogic point of view, the discussion of the nervous
system begins with prana, because in yoga, prana is the main
driver of all of the various processes that happen within the
nervous system. *Prana*, loosely defined, means both "breath"
and "vital energy." In the yogic texts, prana is usually and generally equated with breath, because breath sustains life, and
hence is the vital force or vital energy, and prana and its opposing operation of apana are used to indicate the inhalation and
exhalation. Prana is considered to be just one thing, but it has
different names depending on the job it is performing. I am
only one person, but I am a father to my daughter, husband to
my wife, son to my parents, teacher to my students, and student
to my teacher. In Sanskrit the names for the five pranas are
prana, apana, samana, vyana, and *udana.* In scientific terms,
we say that the nervous system fires in between neurons, sending electrical messages to other neurons through synapses to
direct the body to perform its various operations. In yogic
terms, the electrical synapses are related to the five pranas, and
the anatomical counterpart to the neurons is within the *nadi*
system. In Sanskrit, the five pranas are called the *pancha* (five)
vayus (winds). These five have many different functions associated with each of them. In general, here are the main functions
they perform:

1. Prana rules incoming nourishment.

2. Apana rules outgoing waste.

3. Samana rules assimilation of nourishment.

4. Vyana rules distribution of digested material.

5. Udana rules outward expression.

These five processes apply to all of our interactions with the world, via all the different nervous systems and physiological functions. Applied to the process of breathing, for instance, the breakdown would be:

1. Prana rules inhalation.

2. Apana rules exhalation.

3. Samana is responsible for gas exchange in the lungs.

4. Vyana manages distribution of oxygen to every cell in the body.

5. Udana controls the expression of speech, or outward movement of air other than exhaling (burping, coughing, etc.).

Applied to eating:

1. Prana is ingestion.

2. Apana is removal of waste.

3. Samana is digestion and assimilation.

4. Vyana is distribution of nutrients to the whole body.

5. Udana is the use of energy from food to act and perceive the world.

Applied to experience through the sense organs:

1. Prana is incoming experience.

2. Apana is the release of experience from your mind.

3. Samana is the assimilation of an experience, either positive or negative.

4. Vyana is absorbing an experience into your physical body, as an emotion, feeling, or memory.

5. Udana is acting in the world in response to the way we assimilated or absorbed an experience.

Breathing, digestion, elimination, sensory experience, and engagement with the world: these are some of the functions that the nervous system oversees, and the lists above are some examples of how the five-fold prana operates in order to oversee those same functions. In a more comprehensive way, prana relates not just to breath or energy but to the entire process of nourishment, absorption, elimination, distribution, and expression. Our nervous system is a regulator of those very same processes.

The five pranas operate—or move—through the nadis. *Nadi* in Sanskrit means "a flute, tube, or river." There are three different types of nadi:

1. *Shiras*—the blood vessels
2. *Dhamini*—the branches of the nervous system
3. *Nadi*—the energetic, unseen pathways

In yogic literature, there are said to be seventy-two thousand nadis, a symbolic number, sometimes said to be the signs of the zodiac (twelve) multiplied by the first six chakras; they represent the entire realm of human experience, growth, and emotion. That's where the seventy-two comes from; the thousands are because there are a lot of them. Some texts say there are seventy-two thousand nadis on both the right- and the left-hand sides of the spine. Of these seventy-two thousand nadis, there are ten that are particularly noteworthy. These have their origins below the navel at a place called the *kandasthana*, and end, respectively, at:

* The right and left nostrils
* The ears
* The eyes
* The genitals
* The anus

* The toes

* The tip of the tongue

* And one nadi each for absorbing food and controlling evacuation

There seems to be a correlation between the Western delineation of the nervous system functions and the prana that moves through nadis: they relate to the autonomic functions such as digestion, evacuation, and sexual reproduction, as well as the central nervous system and peripheral operations of the sense organs. They also are connected to balance, a function of the inner ear and the somatic system. Of these ten, there are three nadis that are of primary importance to the yogis: the *ida* (cooling), the *pingala* (heating), and the *sushumna* (singularity). These are also known as the *surya* nadi (sun), the *chandra* nadi (moon), and the *brahma* nadi (pure consciousness).

Conscious breathing practices, such as pranayama, stimulate and balance the energetic aspects of ida and pingala. Through breathing, we connect to our internal world, and when we are connected inwardly, it is easier for us to act in a balanced or responsive way toward the world we live in, our loved ones, our coworkers, political figures that make us want to scream, and so forth. In order for us to live in balance with the world and people around us, our nervous system needs to be in balance, and this balance can be purposefully nurtured. Conscious, steady breathing and breathing through alternate nostrils (see practice B) are both ways of nurturing our inner physiological balance, and are powerful self-regulation tools. On a deeper level, ida and pingala nadis represent duality consciousness, of coming and going, which is also synonymous with the three changing states of consciousness—waking, dreaming, and deep sleep—which we move through every single day. Our identity gets lost in these changing states, in a myriad of constantly changing perceptions, stories, dreams, nightmares, and projections. The sushumna nadi represents the identification we can have with a non-local, interconnected experience of consciousness, of pure being, when the energy that moves back and forth between ida and pingala becomes perfectly harmonized.

The ida (moon) and pingala (sun) nadis end at the left and right nostrils, respectively. They represent all the pairs of opposites, which can be thought of as complementary pairs rather than as opposites:

* Male and female

* Sun and moon

* Logic and intuition

* Hot and cold

* Thinking and feeling

* Inhale and exhale

* Cortisol and dopamine

* Adrenaline and serotonin

* Reactive and responsive

* Sympathetic and parasympathetic nervous systems

The yogis attempted to balance these pairs through postures and, even more, through various breathing practices, such as alternate-nostril breathing and single-nostril breathing. Nostril dominance naturally changes every one and a half to three hours, representing an ultradian rhythm, which is a bodily rhythm that occurs repeatedly through the day in less than a twenty-four-hour cycle. For example, our blood circulation, pulse, heart rate, hormone secretion, and blinking are all ultradian rhythms. A circadian rhythm (from the Latin *circa*, "around," and *diem*, "day"), in contrast, is a cycle that occurs once every twenty-four hours, such as our waking-sleeping cycle. The cycle of nostril dominance is related to both autonomic nervous system innervation of the nasal mucosa and brain hemisphere dominance. In regard to hemisphere dominance, the nostrils are contralateral with the brain, meaning the right hemisphere of the brain connects to the left nostril, and the left hemisphere to the right nostril.[2]

The sides of our bodies are also contralateral with the brain, with the right hemisphere controlling the left-hand side of the body and vice versa. Thus, influencing brain function through breath flow in the nostrils makes perfect sense through the yogis' designation of ida and pingala. Though the pop-psychology hemisphere dominance stereotypes (right-hemisphere-dominant people are more artistic, left-hemisphere people are more intellectual) have largely been debunked, there are certain functions that one hemisphere will rule more than the other. The left hemisphere, for example, is associated with processing language, and the right hemisphere more with interpreting sensory experience. The idea that the left hemisphere is logical and the right hemisphere is experiential has some truth to it, though it is not as simplistic as all that. While the hemispheres do have specialized functions, they also operate more globally than not, communicating through the thick band of neurons called the corpus callosum that connects the right and left hemispheres. For example, the way the brain interprets a math operation will occur in both the right and left hemsipheres of the brain, but memorizing math tables will occur on the left, and estimating a quantity of objects on the right. Both are math operations, but the characteristic of the operation requires different perspectives. Specialization of functions is more related to the way the brain interprets information, and not so much the information itself (like math).[3]

By breathing through alternate nostrils, this particular ultradian rhythm can become harmonized and balanced, as stress, fatigue, and lifestyle can throw off our bodily rhythms (including things like digestion and elimination, which are also rhythms). You can think about breathing through alternate nostrils in the same way that we think about stretching both the right and left sides of our bodies when we do yoga postures; we stretch and strengthen both sides of the body to achieve ease of function and to be adaptable to the environment. When we do that we can have a feeling of harmony and balance in our bodies. In order to actually have that, we also need the ability of both sides of our brain to function, and when we breathe through one nostril at a time, we are enhancing the functionality of our whole brain by focusing on one side at a time, in the same way that we strengthen the right and left sides of the body to make our body function more efficiently as a whole.

There have been several interesting studies on single-nostril breathing that have shown that right-nostril breathing increases cognitive abilities, while left-nostril breathing increases spatial awareness, that is, knowing where you are in relation to what's around you.[4] Spatial awareness allows you to walk through your room at night when the lights are off and not bump into anything, or to walk down the street and sense when you are too close to another person or to the edge of the sidewalk. Though the brain processes information globally, in these studies we still do see the left hemisphere being *associated* with language and math—things that are logical, sequential, rational, analytical, and objective—and with the ways we examine things by looking at the parts. The right hemisphere is associated with things that are less cognitive and more associative, such as poetry, music, and things that are intuitive, holistic, and subjective, or that consider the whole rather than the parts.

The neuroanatomist Jill Bolte Taylor describes in great (and gorgeous) detail in her book, *My Stroke of Insight*, how the left brain thinks and controls walking and talking, and how the right brain feels, is creative, and sees things as a whole, as "trillions of cells sharing a common mind," alive in unity consciousness. While the left brain tries to explain everything, the right brain allows things to be, or perhaps expresses things through music, art, dance, or poetry. By breathing through one nostril at a time, and alternating the flow of breath between them, as the yogis taught, we access our hemispheric brain functions and can balance or influence them with our breath. The innervation of the nasal mucosa will be discussed in the section on the autonomic nervous system.

The nadis, unfortunately, are not all smooth tubes with

energy freely flowing through them—for instance, nadis we can see, like the shiras, are arteries that can get clogged with fats and cholesterol. The subtle nadis, which we cannot see, can get blocked as a result of psychological or emotional issues, and what is impeded is the flow of any one of the five pranas, which may result in a physical problem. A simple example is that when we feel stress or anxiety—which is a psychological reaction to environmental load—our muscles shorten, which can result in neck pain, back pain, or headaches. Western scientists have not found any tissues representing the subtle nadis, though there are some who hypothesize that they may be in the connective tissue and interstitial fluid.

Dr. Neil Theise, of NYU Langone Health, recently published new findings showing that the areas of the body previously thought to be dense connective tissue "are instead interconnected, fluid-filled compartments," making this area—called the interstitium—in effect its own body-wide, interconnected organ, with an ability to communicate to every part of the body. That means that the *lining* of the digestive tract, lungs, urinary system, and surrounding arteries and veins, along with the fascia between muscles, is all interconnected as a bodily organ. Dr. Theise hypothesizes that "the collagen bundles seen in the space are likely to generate electrical current as they bend with the movements of organs and muscles around them, and may play a role in techniques like acupuncture."[5]

Acupuncture is the closest system that has been described to use an unseen network of channels to carry energy to the various body parts and affect bodily processes, emotions, and mental states.[6] This is a very similar idea to that of the nadis and, more specifically, to the *vyana vayu*, which distributes nourishment and messages to the whole body. For example, when an Ayurvedic or Chinese medicine doctor takes your pulse to diagnose your organs and energetic balances and imbalances, he or she is feeling and reading the vyana vayu, which is carrying the pulsations that are emitted from the organs and transmitted through the blood flow. Dr. Theise's findings are more directed toward understanding how diseases like cancer could be spread through these fluid-filled cavities, and toward potential use as a diagnostic tool, and not so much toward modalities like acupuncture. However, the findings point us in an interesting direction, toward understanding the different ways the body communicates with its different parts, and indicate, perhaps, another correlate with the nadi system.

GRANTHIS: THE KNOTS THAT BIND

As the yoga texts say, we have seventy-two thousand nadis, and they all have knots in them; one of the jobs of yoga is to untie these knots. The knots contained in the nadis are called *granthis*; the word *granthi* literally means "a knot." According to yogic wisdom, while our nadi system is filled with knots such as mentioned above, there are three primary knots, the *granthi traya*, that bind us to our limited sense of identification. These are:

1. *Brahma granthi*, which ties us to our survival functions

2. *Vishnu granthi*, which ties us to our emotional bonds

3. *Siva granthi*, which ties us to our intellect, or ties us
    to a limited spiritual identity

We are knotted down to our limited existence by these
three conditions, and by untying the knots, we become free.
When these knots are untied, prana flows freely in all of the
nadis, unhindered by blockages, and can be directed into the
central column of the sushumna. At that point we move from
duality consciousness—that is, breath coming in and going out,
hot and cold, right and wrong, male and female—to unity consciousness, where breathing becomes internal, or situated only
within the spinal column (more on that in the upcoming sections). Pattabhi Jois said that the three granthis are positioned
in between each of the bones of the coccyx, and that through
mula bandha those bones could become loosened and the knots
untied. Other sources state that the Brahma granthi is at
muladhara chakra near the anus, the Vishnu granthi is at the
anahata chakra, at the heart center, and the Siva granthi is at
the ajna chakra, in between the eyebrows. While the granthis
are not visible to the human eye, they are correlated to a location, emotion, and psychological component. Behind these
attributions are the idea that we are knotted down to a limited
view of ourselves through survival necessities, emotional attachments, and intellectual achievements. This is also another
way of looking at asmita as having its basis in the nadi system. The granthis are physical, mental, and emotional mani-
festations of karma that bind us to the stories we tell about
ourselves. Most of our stories can be lumped into one of these
categories:

* Fear for survival, which includes worries about making enough money, finding a partner, or getting old, is the Brahma granthi.

* The need for power, recognition, and our emotional attachments is the Vishnu granthi.

* Intellectual pride, knowledge, spiritual superiority, and self-righteousness are the Siva granthi.

The purification of the nadis leads to the untying of the knots that bind us, which leads to freedom. This in a nutshell is one of the ways that the yogis spoke about the nervous system: as a system of nadis and granthis, powered by prana. One topic that has not been covered here is the chakra system, and that is because the chakras largely correlate to the visceral organs and the glandular system, and my focus here is primarily on the autonomic nervous system and the brain stem. In most of the yogic texts I've referenced, chakras receive very little mention; at most usually only two or three are mentioned, in particular the muladhara, the vishuddi, and the sahasrara (first, fifth, and seventh chakras, respectively).

Now let's look at an overview of how Western science has described the nervous system, and where we might find some correlates.

WHAT IS THE NERVOUS SYSTEM?

The nervous system is the primary communication network that our body is made up of.[7] It contains billions of nerve cells and coordinates a vast array of the body's activities, including automatic functions such as our heartbeat, our conscious ability to control muscle movement, and the integration of different inputs from the sense organs. The "control" or "purification" of the sense organs is actually the way that yoga first was described in the *Katha Upanishad*, dated somewhere between the first and fifth centuries B.C.E.:

Tham yogam iti manyante sthiram indriya dharinam.

||||||

Yoga is considered to be the steady fixing of
the sense organs.[8]

This is largely because the nervous system, which contains our sense organs, constructs and processes the data received through them in order to build the experience we have of the world. For example, our sense of sight: the things we see in the world—every tree, person, object, and animal—enter into our eyes as light photons. These photons refract in the curve of the cornea and enter our retina upside down. The messages the retina receives are sent via the optic nerve to the brain, which somehow reconstructs the incoming data into an image of a world that is not upside down. Each of the sense organs takes incoming information and translates it into experience, but the perception we have of that experience as reality occurs in a little understood relationship between consciousness and the brain. Perception of experience is modified by sense organs and the brain, and we mistake it for reality. Really it is a modification of information, which we perceive in a uniquely human way, and not, for example, in the same way as a cat or a fruit fly.

Each of our nervous systems processes incoming data and internal information, and turns that data into signals that the brain then makes sense of. Some of that data comes from internal organs, like messages sent from the gut to the brain, and some of it is external, like sensing the temperature outside, or that it is night and time to go to sleep.

When I say "each of our nervous systems," it's because we don't have just one nervous system. The main topic of this chapter, and one of the primary concerns of the yogis, is the autonomic nervous system, which is where the automatic functions of the body are controlled, such as respiration, heart rate, blood pressure, digestion, sleep, and body temperature. As a brief overview, these are the main nervous systems of the body:

* The central nervous system is our brain and spinal cord. The brain oversees an unfathomable number of processes, including all of our autonomic functions, our sense of balance, the emotional centers, and our cognitive abilities. Manufacturing and using language, storing memory, experiencing fear or love, and directing many of both the automatic bodily functions, such as heartbeat and respiration, and conscious functions, such as movement, all take place within our brain. The spinal cord is a relay system that carries messages between the brain and the peripheral nervous

system, and messages from the peripheral nervous system to the brain.

* The peripheral nervous system is all of the nervous system tissue outside the brain and spinal cord. It is separated into two sections, the autonomic nervous system and the somatic nervous system. The somatic nervous system is in charge of the voluntary control of bodily movements, including stretch receptors, reflexes, and flexion and extension of the limbs. Reflexes, for example, do not require information to be sent back to the brain. It is the functions of the somatic nervous system that form the larger part of the present-day dialogue regarding asana practice, and what is largely taught in "yoga anatomy" books and courses. The other branch of the peripheral nervous system is called the autonomic nervous system, and itself has two branches, the sympathetic and parasympathetic nervous systems, the main topic of this chapter. The autonomic nervous system oversees all of the automatic functions of the body, such as heart rate, blood pressure, body temperature, digestion, and sexual reproduction, to name a few. Ida and pingala correlate with the functions of these two complementary branches.

* The enteric nervous system is contained in the lining of the intestines. Though the enteric nervous system

is an independent system, it is considered part of the autonomic nervous system, as well as part of an intricate communication network called the gut-brain axis. Through this axis the enteric nervous system (and, some say, the microbiome) interacts with the brain (central nervous system). Marilia Carabotti, a scientist at University Sapienza in Rome, has done extensive research on the gut-brain axis, and has shown that it links the "emotional and cognitive centers of the brain with peripheral intestinal functions," which have a profound influence on mood, digestive health, and higher cognitive functions. It also plays a role in gastrointestinal homeostasis.[9]

❋ In addition to the three main nervous systems, we have sub-networks for specific organs, including the heart. It has a densely packed neuronal network that can send messages and release hormones independent of the brain, though there is not as much research on this as on the other nervous systems. The heart is targeted by sympathetic and parasympathetic nerves, and may have support cells that allow it to function independently from the brain. Indeed, the pacemaker cells of the sino-atrial node, which provide the electrical stimulation for the heart to beat, do not require messages from the brain in order for them to pulse.

One way to think about the various nervous systems is as a massive web of connections linking every single part of your body, every action it performs, every input and output of food or information, and every thought that you have, in a bundle of processes. Thinking and feeling emotions are reflected in the brain as patterns of neurons firing. And our survival functions such as heartbeat and respiration are processes, activities, and, more important, rhythms that guide our days and nights, the years that we pass through, the seasonal changes and relationships we form. We live by the rhythms of wakefulness, of hunger and digestion. The rhythms of our breath change from one nostril to the other, and the rhythms of our breath and heartbeat change with activity and emotions. Our brain-wave patterns fluctuate with our states of consciousness.

All of these rhythms take place within our nervous systems, which is also where most of the underlying mechanisms of yoga, those that make yoga work as powerfully as it does, occur. Nerves send messages, release hormones, and form new connections based on the types of experiences we have, from the time we are babies well into adulthood. They are indeed our internal social network. Yoga practices help to harmonize and bring balance to these rhythms by accessing directly the mainframe of our rhythmicity, the central nervous system, through movement, breath, and focused awareness. As we expose ourselves to yogic ways of living, the sense of expanded awareness that comes with practicing yoga begins to wire itself into our nervous system. We create new, harmonious rhythms, and create synchrony in rhythms that perhaps have been thrown off balance.

The neuroscientist Dr. Bruce Lipton, in his book *The Biology of Belief,* has a succinct definition of the nervous system that I refer to quite often when trying to sum up its interconnected nature and purpose: "The function of the nervous system is to perceive the environment and coordinate the behavior of all the other cells of our vast cellular community."[10] The nervous system perceives, coordinates, and communicates.

Our cellular community is our body and everything in it, about 37.2 trillion cells strong. An unfathomable amount of coordination occurs to accomplish the communication and cooperation of all of these cells—not to mention the coordination of our body with the infinite complexity of the world we live in and that our nervous system responds to. We have to be able to respond not just to day and night, sound, and temperature but also to gravity, the spin of the earth, and our body in relation to the space around us. Our inner ear, the cochlea, for example, is an amazing instrument of both hearing and balance. The cochlea coordinates with balance receptors in our joints and muscles to create something called proprioception, which is how the body knows where all of its limbs are in relation to each other, without us having to look. Imagine if you had to look at your legs with every step you took to make sure they were going in the right place. The sobriety test of stretching your arm out to the side, closing your eyes, and then touching your finger to your nose is a proprioceptive test (drinking impairs proprioception). Asanas strengthen our proprioceptive sense by helping us to notice or develop an interior sense of where our body is in space, and where our limbs are in relation to each other, by virtue of holding ourselves in unusual shapes and angles.

The inner ear also tracks the zero gravity state of the earth's core, so that as the earth spins, we don't fall over.[11] Up-and-down movements, such as in the sun salutations, are stimulating for the cochlea, and thus for our sense of balance. The suprachiasmatic nucleus in the hypothalamus is the pacemaker of our sleep-wake cycle, the circadian rhythm, and tracks the movements of the moon and sun, so that we live by a roughly twenty-four-hour-a-day cycle. To be able to coordinate the almost infinite number of interactions we have with the world around us that occur twenty-four hours a day, for our whole lives, is nothing short of miraculous. This is perhaps why Buckminster Fuller exclaimed, "I am a verb, not a noun!"[12]

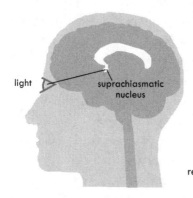

light

suprachiasmatic
nucleus

The suprachiasmatic nucleus sits above the hypothalamus near where the optic nerves cross. It is responsible for controlling the circadian rhythms and regulates many of the body's twenty-four-hour cycles.

Although we are a bundle of processes that participate in the activities of the universe, we somehow feel that we are separate and discrete individuals and that the processes we experience are limited only to us. This is the great illusion of separateness. In actuality, we are a thriving mass of wholeness, interconnected, influential, and capable of instigating change.

Our nervous systems and minds are responsive, reactive, and receptive. Our cells are the biological basis for interconnectedness. They track the movement of the earth and heavens, and help us keep our balance within it.

THE AUTONOMIC NERVOUS SYSTEM: SYMPATHETIC, PARASYMPATHETIC/ VAGUS NERVE

While the functions of the central nervous system oversee both higher-order brain functions and brain stem functions, the properties of the autonomic nervous system are largely survival functions and have been evolving within us for 320 million years. Our brain stem, where our survival functions are housed, literally links us to millions of years of survival-based evolution, and the remnants of those survival traits are still very much within us. While the impulse to survive shows itself through our heartbeat, our digestion of food, our sexual reproduction, and our ability to sense and respond to danger, the higher-order functions reflect the impulse we have to socialize, love, listen, be heard, plan, dream, imagine, and create.

The autonomic nervous system regulates our survival functions. These happen within us automatically, without our having to think about them. Can you imagine going through the day having to consciously beat your heart sixty-five to seventy-two times per minute when you are doing nothing, then consciously beat it faster when you got up to exercise? And at the same time, what if you had to remember to inhale and exhale while you were keeping your heart beating? And on top of that, consciously digest your food, adjust your body temperature to the outside environment, and, if you are pregnant, guide every step of your baby's development in the womb? There is no way we could do even a portion of that, especially because we have to do something called "sleep," and when we do that, we're not in conscious control of anything, even our thoughts. So our autonomic nervous system is really the driving force of our lives. If not for this amazing system, we could not live.

As mentioned, the autonomic nervous system contains two complementary mechanisms, called the sympathetic and para-sympathetic nervous systems. Although the autonomic nervous system is complicated if you dive deeply into the details, the fundamental functions of the various branches can be described relatively simply. I will introduce you to them enough, I hope, to give you an idea of why they are perhaps the most crucial systems affected in yoga.

Looked at most simply, the sympathetic nervous system operates when we move toward activity, for example when we get up in the morning, exercise, or get stressed out. One of its daily jobs is to maintain the body's homeostatic functions by coordinating communication between the organs. The word *sympathetic*—"connection between parts"—was introduced by the physician Galen (129–200 C.E.) in order to explain how the central nervous system communicates with the viscera, or internal organs, including the heart, lungs, and digestive organs.[13] The parasympathetic nervous system operates, for example, when we rest, sleep, digest, or practice mindfulness or relaxation techniques. Our rate of respiration is associated with the balance of these two. The sympathetic nervous system is like an accelerator in a car, and the parasympathetic nervous system, particularly its branches of the vagus nerve, is the brake. In fact, this mechanism is called the vagal brake. It is the vagal brake that restrains the activity of the sympathetic nerves and slows the heart rate down, as mentioned earlier, to between an average of sixty-five to seventy-two beats per minute in a healthy adult. The alternation between accelerator and brake make the heart speed up and slow down with each breath; on the inhalation, the vagal brake releases enough to let the heart speed slightly; with the exhalation, the vagal brake depresses and the heart slows.

The vagal brake, then, becomes an all-important moderator that slows down the fast-moving processes of the sympathetic nerves. Slower respiratory rates can strengthen the vagal brake and thus lead to a reduction of anxiety, stress, and inflammation—a process we will discuss later in this chapter. Faster respiration will occur during activities that require more energetic output, including stress, anger, and illness. According to the biofeedback researcher and psychologist Dr. Richard Gevirtz, anger and anxiety throw off the vagal brake; kindness, appreciation, and gratitude strengthen it.

When we sense danger, and go into a state of arousal or hyperarousal, the sympathetic nervous system kicks into the high-gear state known as "fight or flight." This occurs mainly during hyperarousal, but it is not the only thing that the sympathetic nervous system does, so describing its function only as "fight or flight," which is a primary function of the sympathetic nerves, is not the whole picture of this system. The sympathetic nervous system is in operation at every moment of the day regardless of whether there is danger or not. One of the reasons the sympathetic nervous system is referred to as "fight or flight" is that our brains have been hardwired to learn from what is called a "negativity bias." This means our brain has evolved to learn more quickly from a dangerous or negative experience than from a positive one, and can therefore kick into high gear before we are even conscious of it. As Dr. Rick Hanson communicated to me in an email exchange, "In general, the brain registers negative information and experiences more rapidly than positive information and experiences. In effect, while positive experiences may be more frequent than negative ones, negative experiences generally have more impact." For instance, if on the same day someone says something that we perceive as offensive to us, or gives us a funny look, and someone else holds open a door or pays us a compliment, we might let the slight we felt take on more weight than the compliment or kindness, and ruminate heavily on the perceived insult for days on end.

Because of this bias, we often perceive neutral or non-threatening situations as dangerous, and keep ourselves on high alert much more than we need to be, and this response mechanism is under the jurisdiction of the sympathetic nervous system. This atypical response also occurs from traumatic events. Though the sympathetic nervous system also operates in very basic, ordinary ways during the day, such as when it dilates the eyes or the air passageways of the lungs, we also find many people who are in sympathetic overdrive all the time simply because of their lifestyle, and because of the increasing amount of demands that the world places on us each day. This leads people to remain in a fight-or-flight state when they don't need to be, and this is what is known as stress, more accurately described as *distress* by Hans Selye, the Hungarian-Canadian endocrinologist who gave the word *stress* the associations it has today. One reason that people associate yoga with relaxation and stress reduction is that yoga strengthens, or "up-regulates," the parasympathetic nervous system and slows down or "down-regulates" the sympathetic nervous system, thereby modulating the stress response. "Fight or flight," constant low levels of stress, and sympathetic overdrive all override the vagal brake. Hence, being able to slow respiration can put us back in the driver's seat.

Dr. Hanson suggests that if we deliberately stay with beneficial experiences for a breath or two or longer—ideally while also feeling them in the body and focusing on what is rewarding about them—this will tend to heighten their lasting impact on the brain. In this way, we can increase our learning—including healing, development, and growth—from the experiences we are having. Experience by experience, and synapse by synapse, we can help ourselves acquire greater psychological resources for coping, well-being, effectiveness, and contributing to others. Along the way, we may even be able to sensitize the brain increasingly to beneficial experiences, steepening our growth curve even further.

The parasympathetic nervous system rules the functions that are related to growth, repair, and restoration, as well as "rest and digest." It slows down the heart on the exhalation, and can do so even more effectively through lengthened exhalations. When, through yoga and meditation, the activities of the parasympathetic nervous system are promoted, you can feel everything slow down, and there is an internal calm, a feeling of safety and contentment. Within the parasympathetic nervous system is the large bundle of nerves collectively called the vagus nerve. It makes up 80 percent of the parasympathetic nervous system. The vagus will be discussed in the next section, but it needs to be mentioned at the start of any discussion on the parasympathetic nervous system that the majority of its nerves are the vagus.

It's important to remember that the sympathetic and parasympathetic nervous systems are complementary branches of the autonomic nervous system. They are not opposed to each other. The process of breathing is a good example of a complementary activity. When we breathe, we have to inhale *and* exhale. The sympathetic nervous system moves us toward activity. The parasympathetic nervous system moves us toward rest. When we inhale, the heart speeds up. When we exhale, the heart slows down. We know intuitively that the inhale moves us toward activity because when we need to psych ourselves up for something, like jumping off a high dive, telling someone something difficult, or motivating ourselves to any activity, we often inhale first. The act of inhaling speeds up the heart and allows for greater oxygen exchange as oxygen rich air comes into the lungs.

When we exhale, it is relaxing, calming, and clearing. If we are upset or stressed out and need to calm ourselves down, we typically focus on exhaling, quite often as a sigh or as a long breath through the mouth. As mentioned above, you can think of the sympathetic branch as the accelerator and the parasympathetic branch as the brake of the autonomic nervous system. We speed up, we slow down, sometimes we coast along. Sometimes we have to go super fast, and sometimes we have to come to a quick stop. These two branches control all of that. In yogic terms, the inhalation is associated with prana, or incoming nourishment, and the exhalation is associated with apana, outgoing waste. In terms of the nadis, the heating, active effect is reflected in the pingala nadi, which terminates at the right nostril (heat, remember, occurs because of an increase in speed), and the cooling, slowing effect in the ida nadi, which terminates at the left nostril.

The yoga texts say that by alternate nostril breathing you balance the sun and moon energy of the body. What that means in scientific terms is that you are balancing the hemispheres of your brain and balancing the two branches of the autonomic nervous system; however, the innervation of the nasal mucosa is more intricate then the above listed functions being relegated solely to the right and left nostrils. While the yogis describe the right side as "sun" and left side as "moon," which have correlations to the active potential of the sympathetic nervous system and relaxing potential of the parasympathetic nervous system, the changes in the nostrils are from *both* branches.

The nasal mucosa membrane is innervated by the autonomic nervous system, so that when one nostril is open, it is from sympathetic dominance, and the side that is partially (or fully) blocked is under parasympathetic influence. Approximately every forty-five minutes to three hours they switch sides, and that is what the nasal cycle is. So the right nostril is not solely sympathetic and the left solely parasympathetic; the dominance switches back and forth, which is why alternate breathing is so calming and balancing. However, since the nostrils also are contralateral to the brain, breathing through the right nostril will influence the left hemisphere, and vice versa. There are several levels of influence that occur from single nostril and alternate nostril breathing.[14]

This can be a very useful thing to know when you need to either calm yourself down, prepare yourself for sleep, or increase your energy if you need to focus. The breathing practices can energize you or slow you down. They can heighten your ability to examine things in an analytical way or heighten your ability to move your awareness inwardly to support your innate ability to be contemplative and serene, to sense who you are in an interior way. In Sanskrit, the cooling or moon energy is signified by the syllable *ha*, and the heating or sun energy is signified by the syllable *tha*, which is where the name Hatha Yoga comes from, the union of sun and moon, or sympathetic and parasympathetic. The sympathetic and parasympathetic nervous systems conceptually align with the two complementary branches of practicing asanas: vinyasa and asana sthithi. Vinyasa is the activity and energetic output of the sympathetic nervous system, and asana sthithi is the stillness and absorption of parasympathetic nervous system. These two parts of asana practice, taken together, work to balance the branches of our autonomic functions.

To recap:

* The sympathetic nervous system rules responses to the environment, and processes the metabolic

demands for energy that are placed on us—
everything from running away from a tiger or bear
to getting out of bed in the morning.

* The parasympathetic nervous system rules over
growth, repair, restoration, assimilation, and
relaxation.

* They are complementary systems, and work in
conjunction with each other.

* The sympathetic nervous system is like our accelerator,
and the parasympathetic nervous system, particularly
the vagus nerve, is our braking mechanism.

THE VAGUS NERVE

The vagus nerve is indeed one of the most fascinating and complex nerves of our nervous system. It oversees and connects to almost all of our internal organs, and thus it affects our heart rate and our ability to express emotion through speech and facial expression. It is very likely the most comprehensive communication system we have outside our brain—and this is just a small fraction of what it does. The vagus is the oldest branch of the parasympathetic nervous system, and is the tenth of our twelve cranial nerves. It is the primary nerve of the parasympathetic nervous system, and is actually a large bundle of both sympathetic and parasympathetic nerves, though it is primarily made up of parasympathetic nerves. The cranial nerves come directly from the brain and pass through a special aperture in the skull, whereas the non-cranial nerves of the body pass from the brain through the spinal cord. Most cranial nerves are directed toward one particular function, such as the olfactory nerve, which goes to the nose, or the optic nerve, which goes to the eyes. The vagus nerve, however, travels in what could be called a very comprehensive manner to almost all of our internal organs above and below the diaphragm, hence the name *vagus*, which means "wandering" and is from the same root as *vagabond*. Below the diaphragm, the vagus travels to the stomach, liver, pancreas, bladder, and intestines, and it also innervates the diaphragm muscle. Above the diaphragm, the vagus travels to the soft palate, uvula, larynx, pharynx, heart, and lungs. The vagus originates in two different areas of the brain stem. The part of the vagus that arises from the dorsal motor nucleus controls functions associated with digestion and respiration. The part of the vagus that arises from the nucleus ambiguus, also in the brain stem, is associated with motion, emotion, and communication.

Because the vagus is the longest and most widespread nerve of the autonomic nervous system, its tone is an important aspect of our physiological and emotional health. Vagal tone is like the tone of a muscle. It allows the vagus to perform all the functions that it oversees. Low vagal tone is associated with inflammation in the body, high blood pressure, diabetes, digestive problems, epilepsy, anxiety, depression, and cardiovascular disease. High vagal tone is associated with the reduction of inflammation, better cardiovascular health, high heart rate variability, improved digestion, better sleep, and positive mood regulation. It has been shown that by increasing vagal tone, you can reverse problems that manifest from low vagal tone. Because the vagus nerve is a carrier of messages, if it is not toned and is thus unable to perform its job properly, the body has to compensate in other ways. One of the compensatory responses of the body is to release inflammatory mediators, such as in times of stress, even when they are not needed. This buildup of the inflammatory response can target different organs to cause the problems listed above. When our communication networks are strong, and our internal signaling is clear and direct, the internal organs stay in balance, performing their respective jobs. Dr. Bethany Kok, who conducted research at the Department of Social Neuroscience at the Max Planck Institute for Human Cognitive and Brain Sciences, and conducted her graduate work at the University of North Carolina at Chapel Hill as part of the Positive Emotions and Psychophysiology Lab, has shown that by practicing positive emotion-oriented meditations (see practice C for more on Dr. Kok and how to practice loving-kindness meditation), we can increase vagal tone and physical health, and decrease feelings of anxiety and stress.[15]

Other studies have shown that improved vagal tone can balance blood pressure, and that vagal nerve stimulation, a development of bioelectric medicine, can reduce episodes of epilepsy and instances of rheumatoid arthritis.[16] Practices that strengthen vagal tone strengthen the vagal brake, which is the mechanism that slows our heartbeat on the exhalation and allows the heart to speed slightly on the inhalation. This operation is an indication of cardiovascular health, and also is a snapshot of the health of our autonomic nervous system. Almost every yoga practice—in one way or another—is related to strengthening vagal tone. I'll list many of these practices at the end of this chapter.

DR. STEPHEN PORGES
AND THE POLYVAGAL THEORY

Dr. Stephen Porges, a Distinguished University Scientist at Indiana University and founding director of the Traumatic Stress Research Consortium (who has published over three hundred peer-reviewed papers in addition to many other achievements), has a theory that, through the hierarchical operations of the vagus nerve, the autonomic nervous system responds to real-world challenges—whether a response to danger, or to love—in a predictable way, which means that what the world presents us with, the nervous system knows how to respond to in, hopefully, an appropriate way.[17] He called this the polyvagal theory, and it has quickly become the most influential and inspiring presentation of the vagus nerve and the many aspects of our lives that it influences.

In brief, the theory states that the differences in the structure of the vagus nerve dictate the functions that it performs. The branches above the diaphragm are myelinated nerves, which means the electrical impulses that are transmitted through them can move very quickly, such as messages that can move to the brain in a thousandth of a second. This explains why we can have a rapid change in heart rate when we get frightened or aroused. The branches below the diaphragm are unmyelinated, and therefore their actions are slower. Porges has identified three different response mechanisms of the vagus:

1. *Immobilization:* The "primitive" unmyelinated vagus below the diaphragm into the intestines

fosters digestion, and behaviorally instigates immobilization when faced with dangerous situations by slowing the heart rate and outflow of blood from the heart. This leads to a deep slowing down to protect our energy resources and to protect life. There are also expressions of safe immobilization as well.

2. *Mobilization:* This is instigated by the spinal sympathetic nervous system, which fosters mobilization and the necessary energetic output for "fight or flight." These are myelinated nerve sheaths, so the messages are sent very quickly. You can see this in action when you jump out of the way of a person, a bicycle, or a car (or a bear or a tiger in the woods or jungle in earlier times) before your conscious mind registers that you need to move.

3. *Social engagement and communication:* The mylinated vagus fosters engagement or disengagement with the world around us. This branch of the vagus is related to expression of emotion, breathing, vocalization, and the expression of social behavior and interactions. The myelinated vagus inhibits, or slows, the action of the sympathetic nervous system on the heart, and so it is responsible for promoting calm behavior.[18]

These three parts of the vagus explain much of our relation to the world we live in: engaged and calm social behavior, mobilized protective behavior, and immobilized, defensive, shutdown behavior, such as seen in people who have been severely traumatized or are protecting life. One of the aspects of the polyvagal theory that immediately drew me in was Porges's search to understand the physiological structures that take part in, or affect, psychological and emotional states that we experience. Though not the entirety of his inquiries, much of it can be placed into the categories that I was also interested in: How is it that our physiology affects our mental states, and can we use mental states to affect our physiology? Porges uncovered a world of scientifically verifiable findings to help answer this; all I had was my body, my breath, and some experimentation. The polyvagal theory made a lot of things clearer to me. As a small example, of particular interest to me is Porges's idea of positive immobilization:

> Immobilization, or freezing, is one of our species's most ancient mechanisms of defense. Inhibiting movement slows our metabolism (reducing our need for food) and raises our pain threshold. But in addition to freezing defensively, mammals immobilize themselves for essential prosocial activities, including conception, childbirth, nursing, and the establishment of social bonds.[19]

The practice of asanas, seen in this light, could well be a purposeful immobilization technique for creating a safe, internal environment, for reducing metabolic needs, and for increasing tolerance for pain and discomfort. Indeed, many postures are not comfortable when we first do them, but one of the underlying purposes of yoga is to be able to something that is challenging or uncomfortable in a calm manner, and observe how the mind and the nervous sytem are flexible enough to adapt to challenging situations without being thrown off. The world is not kind, and does not care whether or not we are comfortable or happy. It is up to us to create those conditions ourselves. The ability to stay still and remain observant, sensitive, and calm is an important skill to have when it comes to navigating the world, and the purposeful use of the immobilization response (such as in practicing asanas) may well play an important role in this.

The four neural exercises that give a concrete explanation for several key practices found in the *Yoga Sutras* will be discussed in the coming sections.

THE VAGUS AND EMOTION

Charles Darwin was one of the first to write about the vagus nerve (which he called the pneumogastric nerve) in association with the expression of emotions. He described it as the nerve of emotion, connected from the heart, where emotions are felt, to the vocal cords and face, where emotions are expressed. Darwin saw early on that one of the key areas of influence of the vagus is, as Porges calls it, the heart-brain connection, or, as the Taoists call it, the heart-mind connection. Emotions are processed and expressed because of many different inputs, but the vagus plays an important role in this process. The vagus communicates with nuclei that connect to the cavity of the inner ear called the cochlea, which processes hearing; communicates with the brain to interact with the nuclei that control facial muscles, and thus express emotions; and connects to the larynx to control vocal tone. The combination of speaking, tone, hearing, and facial expression is largely how we not only communicate with each other but are able to identify the type of expression a person is making, whether loving, threatening, consoling, reprimanding, serious, or filled with humor. Interestingly, facial muscles can also elicit changes in brain function and promote changes in emotions, which is why sometimes smiling when we don't feel like it can actually (or at least sometimes) make us feel a little better. Specific facial muscles identified with vagal influence are at the corners of the eyes and mouth, where emotions are largely expressed.[20]

One of the ways or reasons that we express emotions is based on how we are feeling internally. A key distinguishing feature of the vagus is that 80 percent of the vagus nerve is sensory, which means that it receives messages from the internal organs and sends those messages to the brain, to indicate to the brain what the condition of the body is. Those nerves that send messages to the brain are called afferent nerves; the other 20 percent are efferent nerves, which send messages from the brain to the body. You can remember what the efferent nerves are because *efferent* is close to the word *effect*; the efferent nerves carry the instructions from the brain back to the body. The vagus, then, includes bidirectional nerves, carrying messages both to and from the brain, whereas other nerves send messages only in one direction. The vagus nerve informs the brain about what is happening in almost all of the internal organs. For example, when we are breathing in a slow, conscious manner, we are stimulating the vagus nerve endings in the intestines, diaphragm, and lungs through rhythmic movements, which sends a message of calm, safety, and ease to the brain. Many of the practices done in yoga promote and create specific types of messages that are sent through the body to the brain in order to affect our health and engagement with ourselves and the world.

The way the body monitors its internal condition is called interoception. The process of engaging with the outer world, rather than the inner, is called exteroception, and is the function of the sense organs. Much of the pleasure and effectiveness of yoga asanas derives from the interoceptive communication about internal states, so when we twist, bend forward and back, and massage all of the internal organs, we send messages from the internal organs back to the brain, telling the brain things like, "Ah, I am being stretched, I am being strengthened, I am receiving increased blood flow, oxygen, and nutrients . . . I am happy!" Practicing asanas combines both exteroceptive and interoceptive benefits. The ability to increase our conscious ability to sense how we are feeling, physically, emotionally, and spiritually is one of the marks of a contemplative practice. Interoception is a way of being aware of what is happening inside of us.

The practice of pratyahara, the conscious pulling back of the sense organs from their contact with the objects of the world, is another way the yogis worked on the nervous system because exteroception, the outward movement of the senses, uses energy of the sympathetic nervous system for the sense organs to gather information. This energetic drain is either tiring or overexciting for the mind. How can you tell that energy drains from you through your sense organs? Here's one example. After looking at art for two or three hours in a museum, you'll often feel yourself getting tired. Similarly, after two to three hours of binge-watching Netflix you'll feel drained, or as my wife says, brain dead, but your mind will still be whirring about. The metabolic energy used to power the sense organs is used up by outward movement. Control of the sense organs through meditation replenishes our energy because we are not using it up in gathering, grasping, or filtering information from the outside world. Even just sitting quietly and closing your eyes for a few minutes, or for a few breaths, is calming and can replenish your energy. Withdrawal of our awareness away from the outside world through the sense organs, and directing it toward our inner world, is calming for the mind because it down-regulates the sympathetic nervous system.

In terms of emotional expression, the vagus nerve extends to the larynx, where vocal tone is modulated, to the cochlea, where hearing occurs, and communicates with nerves that control the corners of the eyes and mouth, where we display facial expressions to convey our emotions. We can modulate our vocal tone because of the vagus, in order to express kindness, love, affection, anger, or annoyance. When we modulate our vocal tone, the person (or animal) listening can sense what we are conveying through the vibrations that reach the inner ear. The expressions we make with our eyes and mouth further convey the emotional message. That is why, when someone is smiling with their eyes, and not just with their mouth, we feel that they are sincerely smiling. A mouth smile on its own can reflect coldness or hostility. The expression of emotion comes from a wide variety of inputs and confluence of internal processes, the point here being that a toned vagus nerve is going to support and lead to positive, social interactions through being able to read and express emotions in appropriate situations. A person smiling and laughing during a sorrowful funeral, for example, would seem odd. The healthier the state of our internal communication networks, the healthier the function will be of the central nervous system, and this will affect the state of our body. Because many of the processes of the nervous system are bidirectional, by using our body in particular ways—such as practicing asanas, non-harm, and other limbs of yoga—we can influence the state of our internal communication systems: the brain, the vagus nerve, and the way hormones and neurotransmitters are released.[21]

To recap this discussion of the vagus nerve:

* It is the tenth cranial nerve.

* Above the diaphragm it connects to the larynx, pharynx, heart, and lungs, and communicates with nuclei that control facial muscles and the inner ear.

* Below the diaphragm it connects to the stomach, pancreas, liver, spleen, and intestines (as well as the diaphragm).

* High vagal tone is associated with cardiovascular health, stronger immune system function, low levels of inflammation, positive emotions, positive social interactions and behavior, and balanced moods.

* Low vagal tone is associated with higher levels of inflammation, including inflammatory diseases such

as rheumatoid arthritis, cardiovascular disease, epilepsy, and digestive disorders such as irritable bowel syndrome. Low vagal tone is also associated with higher levels of anxiety and depression.

* Vagal tone can be improved through yoga practices and practicing meditations that focus on increasing positive emotions.

* Exhalation slows down the heart by activating the vagal brake.

How can you tell if your vagal tone is high or low? How can you improve its tone and function if need be? Vagal tone is measured through something called heart rate variability. That's what we'll discuss next.

HEART RATE VARIABILITY

Heart rate variability is the beat-to-beat difference in our heart rate, and serves as a reliable measure of the proper functioning of the autonomic nervous system. Remember that when we inhale, our heartbeat speeds up, and when we exhale, the heartbeat slows down. When this happens, it shows that the vagal brake is slowing the heart on the exhalation, and releasing on the inhalation, allowing more blood to move through the heart and increasing the level of oxygenation of the blood via the increased speed of the heart rate. This change in the heart rate is called variability, and one of the many jobs of the autonomic nervous system is to regulate this. Variability is a good thing when it comes to our heartbeat, and in fact to all of our auto-nomic functions. As the psychologist and biofeedback researcher Dr. Richard Gevirtz has said, nature likes a little chaos. Change and variation lead to growth, health, creativity, and expansion. If the heart rate were the same all the time—for instance, if you are flatlining—that would not be a great thing, especially if you want to be alive. Because the heart rate speeds up and slows down with inhalations and exhalations, this activity can be monitored by an electrocardiogram test, which measures the electrical activity of the heart.

Heart rate variability is thus a snapshot of the health and balance of the autonomic nervous system. If the heart does not slow down on the exhale, the vagal brake is not functioning properly, and the sympathetic nervous system is in overdrive, or is turned on but not turning off when it should. This can be triggered by experiencing a traumatic event, or, as in the case of many people, result from exposure to consistent low levels of stress on a daily basis. Our culture right now is one of sympa-thetic overdrive, plugged in all the time: twenty-four-hour news channels, constant use of phones and the internet, texts from our bosses at all hours of the night or on weekends. Since low vagal tone is an indicator of low heart rate variability, we see the same diseases associated with it: heart disease, high blood pres-sure, diabetes, and many inflammatory diseases.

When people are in a constant state of sympathetic arousal, the prefrontal cortex, which is the area of our brain associated with expressing compassion, empathy, long-range planning, strategic thinking, and positive social connections, becomes temporarily impaired as the limbic system enters into a state of hyperarousal. Challenging situations cause the adrenal glands to release adrenaline, and the brain to release cortisol, to help us prepare for a challenge. However, if we respond to every small challenge as though it were a big one, we are constantly releasing these hormones and neurotransmitters faster than our system can reabsorb them. The excess cortisol attaches to receptors in the prefrontal cortex, further impairing our ability to respond thoughtfully as the prefrontal cortex goes "off-line."[22] Cortisol is a hormone that regulates and fights inflammation in the body, but when it is produced in excess, it begins to create inflammation. If we get a cut, fall and get bruised, are exposed to a virus or harmful bacteria, the body will send inflammatory mediators and defense cells to the affected area to heal it, a process otherwise known as the healing response. This is friendly inflammation.

Inflammation that is not well managed by the homeostatic process is considered unfriendly, and can lead to chronic inflammation, joint pain, or metabolic syndrome. Sympathetic nerves that are constantly "turned on" have also been linked to being drivers that can speed the progression of cancer, such as prostate cancer.[23] Consistent low-level stress is one of the drivers of chronic inflammation in society today, and we find ourselves in a vicious cycle—we are increasing the production of the very thing that is supposed to help suppress inflammation and fight harmful bacteria, creating conditions for its overproduction because of our overstimulated lifestyles. High heart rate variability, then, becomes a measure that will show us how well the vagal brake is functioning, and how well we can turn off the stress response when we need to, and the tone of the vagus nerve plays an important role in mediating inflammation. There are many apps available that can give you a basic measure of your HRV, so that you can monitor it from day to day and get an idea of where your stressors are coming from. Professional athletes are turning more and more to HRV monitors for heightened performance, because they can ascertain the best time for training and, more important, recovery time, depending on HRV levels.

BRAIN STEM FUNCTIONS

The brain stem has three different sections: the mid-brain, the pons, and the medulla oblongata. The different parts of the brain stem control the steady and constant stream of messages that pass between the brain and body, as well as rule the functions of the autonomic nervous system, such as respiration, heart rate, blood pressure, digestion, sexual reproduction, and body temperature. These survival functions are also related to the body's homeostatic functions, which control how the body maintains balance, adapting throughout the day to the demands of the environment in big and small ways. Homeostasis is controlled by the hypothalamus and the neuroendocrine system, and includes blood pH balance (which has to do with oxygen and carbon dioxide ratios), core temperature, blood glucose levels, arterial blood pressure, and the balanced composition of many of the chemicals that make up our body. When the temperature changes outside, our body regulates itself to balance out that change; when we are exercising, the heart rate speeds up to maintain the increased demand our cells have for oxygen; when we are digesting, blood flows toward the digestive organs to support assimilation and absorption, drawing blood away from the brain (this is why you feel drowsy or unable to think clearly after a big meal). Maintaining homeostasis uses a lot of energy—we don't just get into balance and stay there; it's a constant coming and going. The same is true for yoga poses. We never come into perfect balance and just stay there, we are always micro-adjusting. Balance is really the act of balancing.

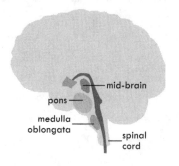

The brain stem is made up of the mid-brain, the pons, and the medulla oblongata. Together they control the constant and steady stream of information between the brain and the body.

While there are overlaps in brain stem functions, there are some specific jobs that each portion has, too. The pons rules the processes of sleep, bladder control, equilibrium, taste, eye movement, and posture; it also determines our breaths-per-minute rate and amount of air breathed. The conscious slowing of the breath indeed allows us to purposefully manipulate brain function via the pons. The mid-brain is associated with visual and auditory processing, sleep and wake cycles, alertness, and temperature regulation. The medulla oblongata rules autonomic functions such as respiration, heart rate, blood pressure, and bodily reflexes such as sneezing and vomiting. Practices such as asanas, conscious breathing, and drishti are targeting many of the processes that the brain stem oversees, such as breathing rate, balance, equilibrium, and visual processing. All of the basic yogic practices of tristhana directly exercise, strengthten, and balance neural functions.

So what is the brain stem in relation to the rest of the brain? The brain stem rests between the spinal cord and the entrance to the higher functions of the brain; it is a gateway through which information passes from the body to the brain. It is the oldest part of our brain, and contains within it the vestiges and imprints of 320 million years of evolution. We share brain stem functions with all other mammals. Our limbic system, which processes emotions, memory, and sense of balance, developed about 100 million years ago. The neocortex is the most recent evolution of our brain, only about 60,000 years old, and developed in response to our species becoming increasingly social animals. The neocortex is truly our social networking system.

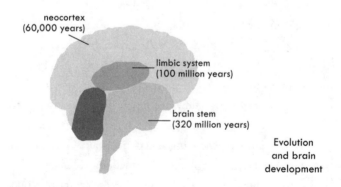

neocortex
(60,000 years)

limbic system
(100 million years)

brain stem
(320 million years)

Evolution
and brain
development

It is interesting that many of the practices that the yogis did directly addressed the survival functions, those functions that are processed through the brain stem. They restricted breathing and breath rate through pranayama; they controlled the heartbeat and blood pressure through bandhas; they controlled hunger through fasting; they practiced celibacy to control the sexual drive; and, most popularly, they practiced postures to improve equilibrium, balance, and levels of alertness through mindful movement. Hariharananda mentions in his commentary on tapas that the functions of the brain stem are natural. But to transcend nature, and to become supernatural, the yogis practiced the opposite, and developed techniques for controlling those natural functions.[24] Why could this be?

KLESHAS IN THE BRAIN STEM

My supposition about this is quite simple and straightforward: I believe that the physiological basis for the kleshas, the obstructions to knowing who we are, which we discussed in chapter 7, exist or have their basis in the brain stem, in our survival functions. The final klesha is abhinivesha, which is fear of extinction. Abhinivesha is not a fear that is formed through conscious thought; it is a fear that somehow exists in the very life of a cell. The cell does not necessarily identify fear, but there's an impulse that moves the cell to avoid danger, to seek nourishment, and to breathe in the way that only cells do. We, too, have that impulse as the basis for the unconscious functioning of our nervous system. However, as humans, we also are aware of it on a conscious level. We fear extinction and cling to life both through our automatic survival functions and through using the sense of fear to guide us away from dangerous situations.

This began happening not when we were born, but billions of years ago as life developed on our planet, and the impulse to live and to survive forms that basis for our autonomic functions. We cling to our breath, our heartbeat, our need to eat, and our need to reproduce. Clinging to life is not a philosophical problem; it is a physiological imperative. However, this clinging to life, through the survival functions, leads us to cling to a false sense of a separate self, and the stories we construct about this self, which are reflected through asmita. Both asmita and abhinivesha have a physical location in the brain stem, in the area where our survival functions are located. But this is encouraging, because anything that is reflected through a physiological process can be worked with, and if we can locate it, and work with it, we can change it.

I can imagine the yogis asking, What would happen if we transcended the physiological mechanism that ties us to a false sense of self? What would happen if for a little while each day we controlled, slowed, or even stopped our heartbeat, our need to eat, to have sex, or even to breathe? Who would we be if we lessened our identification or became completely comfortable with the survival functions for even a short period of time, so that fear of survival did not rule us? Would it be possible to use our bodies to transcend the brain functions that are tying us, like granthis, to our limited physical form? That bind us to our limited, constructed sense of personality, the sense of "I-ness" that arises along with this clinging to life? It would mean that we could work with the root of asmita, of "I-ness," where our constructed narrative grows from. After that, perhaps, we could move toward inhabiting higher levels of the brain processes and awareness, starting with the limbic system, where we could address balance and spatial awareness, our place in the world. We could thin the walls of the amygdala, where fear is processed, and then purify our memory and emotions processed through the hippocampus and the hypothalamus. From there, we could pass even deeper into the realm of all-encompassing compassion processed through the prefrontal cortex, and then enter into the unity consciousness of whole-brain functions, which occurs when we experience bliss or transcendent awareness. *This is not to say that our biology, or our brain, creates consciousness; it is to suggest that we can use our biology and brain to access higher states of consciousness.*

I am not trying to give an answer to what consciousness is, or where it comes from; I am simply showing how we can use the physical materials we have to work with to experience higher meaning and purpose. It is a suggestion that uses the practices of the yogis as a jumping-off point for the experience of higher brain functions that can lead to self-knowing, or transcendent states of being, or happiness. Each level of our brain functions processes aspects of our lives that correlate to a spiritual journey toward self-discovery and freedom. The subtler the practices become, such as moving from asanas into pranayama, and then on to compassion practices, the more our deeper brain structures and nervous systems are affected. When we feel safe and grounded, the circuits of the brain stem are not directing us to be in a defensive mode, as when we perceive danger or are anxious. When we perceive danger, the circuits of the brain stem turn on, and the circuits that allow us to access higher brain functions are turned off. Because the brain stem mainly rules autonomic functions, the diversity of experience is limited to survival issues. However, when survival issues are not ruling us, the neural circuits that communicate with higher brain functions are open, and we can enter into an infinite world of a diversity of experience that comes along with the limbic system and cortical functions of the brain. We have higher brain functions because they are meant to be used by us, but we can't use them when we live in fear.

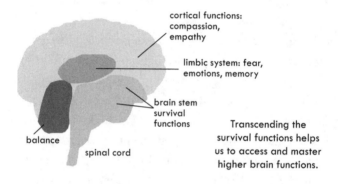

cortical functions: compassion, empathy

limbic system: fear, emotions, memory

brain stem survival functions

balance

spinal cord

Transcending the survival functions helps us to access and master higher brain functions.

THE PRACTICES OF KRIYA YOGA

So what practices thin the kleshas, so that the obstructions to knowing who we are become porous, and the light of self-knowledge shines in our awareness? When we speak of asmita and abhinivesha as having a physical location in the brain stem, what can we do to weaken their hold on us? These are the practices of kriya yoga: tapas, svadhyaya, and Ishvara pranidhana, and they are the gateway, or indirect practices, that work on moving us toward higher states of yoga.

Tapas means "to cook," or "to heat," and includes practices such as postures, pranayama, and meditation. Tapas works directly on restraining the survival functions of the brain stem by controlling respiration and heart rate through asanas and pranayama; digestion through dietary restriction and healthy eating habits; hunger, thirst, body temperature, and blood pressure through meditation and certain breathing practices; and the sex drive through lifestyle choices. These practices are called tapas because they cause a little bit of discomfort. The texts say that if we get upset by small things, then our minds are not ready for the deeper levels of yoga. But by building up endurance through purposefully engaging in small amounts of hardship—such as calmly staying in a challenging pose for some time, or sitting for periods of meditation without moving—we build up the strength of mind to be able to endure discomfort, and this helps to make the mind steady. Life will always be filled with hardships, and to be able to endure them patiently is the sign of a yogi.

Svadhyaya is chanting mantras, and the study of spiritual texts. Chanting works on the emotions through devotion, and as well accesses abstract thinking mechanisms of the right hemisphere of the brain through non-symbolic sounds, such as mantras. The right hemisphere of the brain is dominant in the regulation of autonomic functions, and thus is associated with the expression and interpretation of emotion.[25] Love and devotion that are generated through devotional chanting address the limbic system, and reach higher-level functions than survival mechanisms. These things taken together make devotional chanting another mechanism for regulating autonomic functions *and* expressing love or ecstatic, transcendent emotion. As well, in order to chant, you also have to regulate your breathing, so at the same time that it is a devotional practice, it also is a tapas because you have to use breath control.

Chanting is particularly interesting because music and rhythmicity of sound are among the few things that affect whole-brain function at the same time. Most things we do stimulate either the right or the left brain, or the brain stem or prefrontal cortex; our activities or thought processes are intellectual or intuitive for example, but very infrequently does the whole brain respond in a coherent fashion. Concetta Tomaino, DA MT-BC, a music therapist and the executive director of the Institute for Music and Neurologic Function, relates that music stimulates whole-brain function simultaneously, including tone, pattern, rhythm, meaning, and memory. This leads to whole-brain and whole-being functions. Music helps to reinforce memory and experience, and is why when we hear songs that are important to us, it recalls in us not only a particular time of our lives but also *the feelings* associated with that time, and so it useful in the treatment of Alzheimer's and in the rehabilitation of stroke victims. TM mantra meditation has been shown in scientific studies to also make the synapses of the brain fire in a coherent pattern.

Our brain is a pattern seeker, and we are wired for patterns associated with rhyme, rhythm, movement, and emotion. Engaging in these patterns wires our brain through neuroplasticity, the ability of the neurons of our brain to wire, rewire, grow (through neurogenesis), and create new connections. As well, we have built-in rhythms in our body, such as our heartbeat, pulse, the blinking of our eyes, and respiration—rhythms that we can listen to when we meditate, and rhythms by which we live. A newborn rests its head against a mother's breast to listen to her heartbeat for comfort and safety, recalling the rhythmicity of the heartbeat it nestled next to during the nine months of gestation.

Further, all language is based on rhythm, pace, pitch, tone, and cadence—or what is called prosody. The ancient *Taittiriya Upanishad* lists the rules of chanting and of speech in its opening paragraph, and states that the rules of prosody lead to the understanding of all of the ways that consciousness and matter have linked together to create all manifestation, listing the five perceptible objects as the universe, light, learning, progeny, and self (a self based on articulation through speech). Clearly, chanting and singing are among the best ways we can encourage whole-brain patterns and functions. Music and chanting literally wire our brain for deeper levels of experience and understanding. They can also help heal trauma, the aftereffects of strokes, and memory retrieval. The body is not only a bundle of processes; we are also a collection of rhythms. Much of yoga is drawing these rhythms together into a synchrony, because there are so many things that happen during the day, and in our lives, that cause us to be in a dis-synchrony, excess stress being foremost among them.

Ishvara pranidhana means either "surrender to God," if you are theistic, or "surrender to the unknown," if you are not. Surrender moves us toward the highest levels of brain functions in the prefrontal cortex. Relationships are the jurisdiction of this part of the brain, so our relationship with the divine or with nature will be processed in this part of the brain. Built in to Ishvara pranidhana is the sense of not-knowing, of not having to "figure it all out," and that it is okay to let things be, that we don't have to control everything all the time. This creates a space within us where we can become a little more accepting, receptive, and forgiving. In the yoga texts, the perfection of Ishvara pranidhana is said to lead to unity consciousness, transcending the limited sense of "I." This occurs through the clarity of insight of the awakened mind. With Ishvara pranidhana we arrive back at the idea of relation, as in the relationship with the divine, or with nature, returning us to the definition of *yoga*, given in chapter 1 as "relation." Relation is a physiological imperative, a transcendent principle, and the foundation of our being integrated and interconnected with the people and natural world that we are interdependent with.

The practices of kriya yoga (tapas, svadhyaya and Ishvara pranidhana) mirror the triune brain theory, which was created by neuroscientist Paul MacLean, who first described the parts of the brain in regard to evolutionary theory. Each part has its own functions, but they are not independent; rather, they influence and complement the other functions. The reptilian brain is the brain stem and the cerebellum; it rules the survival functions and balance. The mammalian brain, called the limbic system, is where memory is stored and emotions are processed. It's composed of the hypothalamus, the amygdala, and the hippocampus. The amygdala is responsible for processing fear, and when we react to fear, the brain stem heats up the body (such as sweaty palms or a flushed face) and speeds the heart rate. The hippocampus is involved in memory retention, and the hypothalamus coordinates many of the activities of the autonomic nervous system, including homeostasis. The neocortex, or neomammalian brain, was the last part of the brain to develop and is where language has developed, as well as our capacity for abstract thinking, strategic and long-range planning, the expression of compassion and empathy, and an almost unlimited capacity for learning and creativity.

Tapas, svadhyaya, and Ishvara pranidhana fit in perfectly with the triune brain theory, and while the three practices can theoretically be done independently, they are as intertwined and influential over each other as brain functions are. These practices remove the obstructions that cover our awareness and lead to the ability to have a measure of conscious control over the survival functions in the brain stem (tapas), the purification and expression of the emotions in the limbic system (svadhyaya), and finally the development of love, compassion, empathy, and positive social integration in the prefrontal cortex (Ishvara pranidhana). When these three aspects of brain function are balanced, it leads to whole-brain functioning, rather than the fractured functioning that we experience on a daily basis due to stress and living out of tune with nature. When we experience whole-brain coherence, we are in a state of connectivity, self-knowing, and total integration. It is the most relaxed inner state of being that we can experience on a conscious level, the state described as transcendence by the yogis, nirvana by the Buddhists, and rapture by the mystics.

To recap, tapas is related to physical practices, svadhyaya is verbal and emotional, and Ishvara pranidhana is mental and emotional. The physical practices target the brain stem through asanas, pranayama, and restrictions. The verbal and emotional practices target the mid-brain, made up of the centers that process fear and memory, and that support homeostatic functions.

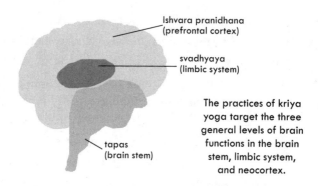

Ishvara pranidhana
(prefrontal cortex)

svadhyaya
(limbic system)

The practices of kriya yoga target the three general levels of brain functions in the brain stem, limbic system, and neocortex.

tapas
(brain stem)

Kriya Yoga and Its Correlates

tapas	physical/restrictions	asanas/pranayama	brain stem
svadhyaya	verbal/emotional	chanting/bhavana	limbic system
Ishvara pranidhana	mental/emotional	surrender/compassion	prefrontal cortex

Ishvara pranidhana brings us into the higher-brain cortical functions, where we express compassion, empathy, and connectedness, as well as strategic thinking and planning. This trifecta of practices supports whole-brain functions, and leads us to express our fullest range of human potential and existence, which includes the ability to step outside of our limited views about ourselves and the world, and experience a positive, transcendent, and inclusive perspective.

NEURAL EXERCISES

Stephen Porges has shown how important the tone of the vagus nerve is to our ability to self-regulate, have healthy relationships, and have a balanced, functioning nervous system that maintains all its homeostatic processes properly.[26] In examining religious traditions and rituals traditions, he identified four distinct categories of practices that all point to vagal nerve regulation via strengthening vagal tone.[27] These neural exercises draw together almost all of the material that we have covered so far in this book, and reveal in Western, scientific language a sound physiological basis for all of the main yogic practices of Ashtanga Yoga, and for the practices of kriya yoga in particular. The neural exercises, as explained by Porges, are:

* *Posture:* Change of posture, even just sitting up straight, stimulates the baroreflex in the carotid artery, activating the baroreceptor nerves that monitor and control blood pressure and lead to mood change. Posture is used in all of the world's religious and mystical traditions, perhaps for the very reason of mood change. In Islam, prayer is based on prostrations; Hindus practice asanas and prostrations, which form the basis for sun salutations; Hasidic Jews rock back and forth during prayer (davening); Sufis—in particular the whirling dervishes—twirl, and ecstatic dance is a type of bodily posture. Swaying, rocking, prostrating, spinning, and bowing: all of these stimulate the vagus nerve through different pathways. These are also ways that infants and children move naturally as they go through the developmental stages: rocking back and forth,

swaying in circles, rolling on the ground, crawling, and finally walking.

* *Breathing:* Respiration affects the abdominal afferent nerves through the movement of the diaphragm and belly, and these nerves send messages to the brain of rhythmicity, safety, and contentment. Respiration also directly stimulates and tones the vagus nerve through stimulating the larynx and the vagus nerves that innervate it. Ujjayi pranayama, which is performed by lightly tightening the glottis and breathing with a whispering sound, and then exhaling out of either the left or right nostril, stimulates the vagus through massaging the larynx and down-regulating the sympathetic nervous system. The audible breathing heard in some yoga practices, which is sometimes called "ocean breath," is also tonifying for the vagus nerve (ocean breath is sometimes used synonymously with ujjayi, but the two are separate practices).

   Breathing deserves a special mention here because of its ability to down-regulate the stress response that occurs in the sympathetic nervous system. Breathing, especially lengthened exhalations, can be both voluntary and involuntary, and has the benefit of turning on the vagal brake. All of these exercises taken together—but especially the breathing aspect on its own—bring us into the present, which is a non-defensive state. As Porges says, when we are non-defensive we are connected. When we are in sympathetic arousal and are in a defensive mode, we are disconnected, isolated, and closed or locked into survival mode. As human beings, we cannot exist without each other. It is not just that we need to feel connectivity; it is that we are part of a collective already. Separation is an illusion. By denying our innate connectivity, we divorce ourselves from our collective reality, our collective consciousness, and suffer for it. Many of our preventable, non-communicable diseases, such as anxiety, depression, heart disease, and some cancers, are caused by biomarkers for inflammation that are activated by living in sympathetic arousal, by living in a defensive state of being.

❋ *Vocalization:* Chanting and singing stimulate the laryngeal nerves, which are connected to the vagus nerve, and also stimulate whole-brain function. Vocalization stimulates the vagus nerve as it runs next to the larynx; chanting and singing in particular have a massaging effect on the vagus nerve, the gag reflex, and also the expression of emotion. With vocalization, we are making not just sounds but sounds that have meaning and feeling behind how something is said. Many years ago one of my students suffered a stroke and lost her gag reflex. While in the hospital, she began listening to, and then chanting, the *bija* mantras of the chakras, and regained her gag reflex within days.

❋ *Behavior:* The heart-brain axis is affected by our behavior, and positive emotions and thought processes such as kindness and appreciation increase vagal tone through the heart-brain axis. Behavior such as acting with love, kindness, compassion, and care has also been shown to tone the vagus nerve. The study by Bethany Kok referenced earlier has shown that the practice of loving-kindness meditation has the same toning effect on the vagus nerve as vagal nerve stimulation, the surgical implant of magnets in the throat used for stimulating vagal function.

Simply from reading this list, you probably have grasped where this leads. It's striking that these neural exercises align completely with the practices we do in yoga:

❋ *Posture:* as in the many yoga asanas, used during both yoga practice and meditation

❋ *Breathing:* pranayama, and the conscious breathing done in postures, as well as awareness of breath used in meditation practices (neural exercises of postures and breathing are both related to tapas)

❋ *Vocalizations:* the chanting of mantras in svadhyaya, such as the chanting of *om* or other mantras; Vedic chanting; the call-and-response singing of kirtan

❋ *Behavior:* the observances in the yamas and
niyamas, and also the loving-kindness practices
such as friendliness, compassion, sympathetic joy,
and equanimity of mind spoken of in the *Yoga
Sutras*, in chapter 1, verse 33 (see practice C, on
loving-kindness meditation). Behavior can also be
seen in the devotional mood of surrender, or Ishvara
pranidhana.

It is clear that all of the exercises that Porges lists, which
have specific and measurable effects on the vagus nerve and
strengthening vagal tone, are exactly the same as the fundamental yoga practices that for thousands of years have been used to
create a mental and emotional state of openness, compassion,
and adaptability. Strengthened vagal tone leads to self-regulation,
and the ability to self-regulate can lead to one's ability to have a
measure of control over the autonomic functions. Regardless
of whether or not you are not interested in the deeper levels of
these practices, even on a basic level they will lead to us becoming more fully integrated human beings. There is no doubt in
my mind that this correlation between the vagus nerve and yoga
is purposeful, as these four categories were all things the yogis
did intentionally because they knew that this was how they
could control their nervous systems and learn to self-regulate in
order to thin the cloudy veil of the kleshas that cover unbounded
consciousness.

We have one more piece of the puzzle to fit in: the concept
of *kundalini*.

KUNDALINI

Kundalini is the name ascribed to the creative energy of the universe that lies latent within each of us. When all the impurities are cleansed from the nervous system, the granthis untied, prana and apana in an equilibrium, and the mind controlled, this energy, which lies dormant at the base of the spine, is said to wake up and rise up through the sushumna nadi until it reaches the skull and enlightenment occurs. In some texts, such as the *Yoga Yajnavalkya*, kundalini is called a blockage that is at the entrance of the sushumna.[28] When this blockage is removed, prana flows up and down the sushumna, rather than in and out the nostrils.

Kundalini is not mentioned in the *Yoga Sutras*. The point of this book is not to delve too deeply into mystical ideas that are unverifiable through sources other than testimony. However, I bring up the subject now because there are precisely two sources that I know of that identify the sushumna nadi as existing in the vagus nerve complex, specifically as it rises from the heart to the brain. The first source is Pattabhi Jois, who related this to me in conversation, and the second is a small book called *The Mysterious Kundalini*, by Dr. Vasant G. Rele of Mumbai, written in 1927. In his book, Rele provides careful descriptions of both yogic practices and the nervous system, and identifies the chakras and nadis with their scientific counterparts. He explains that by controlling both the afferent and efferent vagus nerve endings at the solar plexus and in the brain stem—the nerves that convey information to and from the brain—the yogi can gain control over the autonomic functions.

Rele says, "This is what a Yogi desires, so that the normal automatic function may not interfere with his desire of becoming one with Him who is all pervading."[29] The sixth paragraph of the *Taittiriya Upanishad* also locates the sushumna as rising from the space of the heart and, cryptically, into the "two arteries of the upper palate" behind the uvula, and then rising into the bones of the skull. The vagus nerve indeed innervates the soft palate, uvula, and root of the tongue, so it is a likely correlate.

According to the yogis, when we breathe in and out through the nostrils, we are in the world of duality, of coming and going, of all the complementary pairs of opposites. When prana moves into the central channel of the sushumna, connecting from the heart to the brain, breathing is said to occur internally and we exist in a state of non-dual unity consciousness.[30] When awareness becomes absorbed inwardly, the breath ceases to move in or out. As the *Hatha Pradipka* says, when the breath stops moving, the mind stops moving and rests in stillness. This is called kevala kumbhaka, or the natural ceasing of the breath. It occurs not from holding the breath, but from the breath *ceasing*. An example of this is when you are in a deep state of concentration, and notice that you haven't been breathing for a short time, but you haven't been holding your breath, either. By definition, kevala kumbhaka is when respiration occurs within the spinal cord: it is one of the pinnacles of pranayama practice.

We do, in fact, have respiratory centers in the brain and spinal cord that control respiration independent of the lungs, heart, and diaphragm, called the central pattern generators. They produce rhythmic internal patterns of respiration even when there is no breath coming in, meaning that even if you are not breathing physically, the rhythmicity of respiration will still occur inside you, and keep you alive. As well, there is another cluster of interneurons called the pre-Bötzinger complex, which is in the medulla, along with our survival functions, and is essential for generating respiratory rhythms in mammals.[31] Murali Doraiswamy, a professor at Duke University and a leading researcher at the Duke Institute for Brain Sciences, related to me that "there are many intrinsic pacemakers in our body, including many that survive even after a human is declared medically dead. The neurons in the suprachiasmatic nucleus in the hypothalamus are one example, and the brain stem is another." (The suprachiasmatic nucleus is the cluster of neurons in the brain we spoke about earlier, which controls the circadian rhythm.)

The exact mechanism for how this works is still not understood. The pre-Bötzinger complex is made up of both pacemaker and non-pacemaker neurons, which means that coded within their cells is a pattern that will occur independently of central nervous system stimulation. The heart cells have the same makeup: if you remove a heart cell from the sinoatrial node, which is the pacemaker of the heart, that cell will continue to beat. The heart is made up of cells that individually beat on their own, or propagate their own electrical impulses, in at least three nodes. The cells of the pre-Bötzinger complex do the same thing, but for respiration—they don't depend on air coming in and out of the lungs in order for them to propagate their respiratory pattern.

In the early studies of yogis who claimed they could stop their heartbeat, EKGs and other measurements showed that the beating of the heart—that is, the pacemaker—did not stop, but somehow the flow of blood through the heart did slow or was momentarily interrupted. Based on anatomical structure, it's a good guess that the same basic thing could occur with internal breathing, whereby external flow of breath ceases, but the internal pacemakers continue to do their job of sustaining life.

Doraiswamy further elaborated that there are some scientists who would say that every single cell in our body contains its own pacemaker or rhythm generator, which is a mind-boggling proposition. William Bushell, Ph.D., a Massachusetts Institute of Technology, Harvard University, and Columbia University affiliate, has suggested that yogis are also able to activate the cellular state of hibernation when they enter into deep states of samadhi, and greatly slow or suspend bodily functions during that time.[32] According to at least one study that Doraiswamy shared with me, hibernation-type states can induce changes in the medullary pacemaker that make them more resilient. The study also showed that during hibernation states, the electrical activity of the cortex and other higher brain functions becomes silent, yet the respiratory pacemakers continue to function. Advanced yoga practices such as kevala kumbhaka may allow the yogi to temporarily stop autonomic functions such as breathing or blood flow to the heart, and yet still remain alive, by entering into a hibernation-like state.[33]

The four neural exercises identified by Stephen Porges occur in every mystical and religious tradition to some degree, not just yoga. Through these four practices, or exercises, we can directly access the inner automatic functions of our physiology and begin to have a measure of control over our mind and emotions, so that we are not simply at the whim of our autonomic nervous system. This is not to say we want to overpower our autonomic nervous system all the time; it means that in the same way we can influence our health through diet, and our energy through sleep, exercise, and perhaps having one or two hobbies, we can exercise a level of direction over our nervous system through yogic practices, and this direction can lead us toward greater levels of health, happiness, and fulfillment of purpose. At their pinnacle, they lead to utter and complete inner stillness.

CONCLUSION

There is of course a huge amount to be said on many of the points I've presented so far in this book. I've tried to cover and include much of what I have learned, as well as some of my own thoughts about the intersection and influence of yoga on the nervous system. Other authors and scientists will no doubt fill in the holes I've left, correct my mistakes, and add to the corpus of insights as time goes by. Regardless, we can continue to count on at least two points I've presented to remain true. First, we know that high vagal tone lowers inflammation, improves resiliency, increases adaptability, supports homeostasis, and helps to control mood and emotion, and that the practices within yoga (kriya yoga) tone the vagus nerve. All of these things contribute to a healthier, happier, more connected life. When we have too much inflammation in the body or mind, everything is thrown off, especially our moods and emotions. And indeed, we would like to be happy and kind and effective human beings, so keeping the vagus nerve toned is a gateway practice for other positive things to manifest within us.

Second, many yoga practices address the autonomic nervous system, where our survival functions are processed and regulated. Our survival functions keep us alive and identified, on a psychological level, with being alive. Because every action has an equal and opposite reaction, the other side of the survival coin is being dead, so avoidance of death is hardwired into surviving—for in order to survive, we cannot be dead. Hence, the survival functions help us avoid death, and the avoidance or fear of death becomes subtly and intricately bound into our survival, which is abhinivesha.

The identification with our individual life is where our false or limited personal narrative, or asmita, develops. Asmita is an offshoot of avidya, of not knowing who we are, and the attachments we form based on our likes and dislikes are an integral part of our narrative. These are the obstructions to knowing who we are, and they have a physiological basis in the brain stem that can be influenced via kriya yoga, or neural exercises, as Porges calls them. Happiness is said to manifest in our minds and hearts spontaneously when we know who we are, and when we are free from fear. It follows, then, that the practices of yoga provide a way for us to be happy not just as an idea, but as a physiological certainty. Seen in this light, we can shift some of the emphasis away from current ideas that yoga is about difficult poses, that you need to be flexible in order to do it, that you need to buy special clothes to wear and an expensive yoga mat, or that it is about controlling the mind and not having thoughts. This approach, as the molecular biologist Alexandra Seidenstein says, is a body-over-mind approach. We train the body to affect the mind to operate in the way that we want it to. Many who practice yoga find that by doing something physical, they are able to work out problems that are bothering them in their mental sphere. Why is it that a physical practice can help solve something that is seemingly mental? It's because many of the things that occur in our minds, including emotions, have a physical basis or counterpart that when worked with can release emotional or stressful mental states, such as stretching or applying deep pressure to the muscles, which are both releasing and grounding for the nervous system. Emotions and thoughts are contained in our body, because the mind and body are not separate; they are a continuum. The models presented in this book, such as asmita having a physiological basis in the brain stem, are supportive of that basic idea.

So what are the things we know that we can do to improve vagal tone and activate the parasympathetic nervous system? In regard to daily practice we can:

* Practice asanas, pranayama, and meditation with awareness.
* Practice coherence or resonance breathing.
* Chant mantras and sing.
* Be kind to people.
* Hold positive and loving thoughts about ourselves and others in our mind.

In regard to lifestyle changes, we can:

* Introduce small changes that will help us lead a balanced life.

* Allow ourselves to experience joy and pleasure in our practices. They should not be a chore or an obligation, but a choice, something that, even when a little difficult or tiring, does not weigh heavily on us.

* Find ways to minimize stress, perhaps in the form of taking short breaks during the day.

* Make sure we get enough sleep.

* Include foods in our diets that will contribute to the health of the microbiome, thus sending positive messages to the brain.

And on a spiritual level, we can:

* Find ways to reframe our perspective on life to see the world, and our place in it, as part of a collective.

* Try to sense during the day the interconnected nature of existence, how the air we breathe, the light we see, and the earth we walk on are extensions of our being.

* Notice how all these things together evoke a feeling of gratitude and appreciation, and a sense of a shared existence with all other beings. This is truly where inner joy and fulfillment of purpose come from.

This one simple thing called yoga leads to the loosening of the narratives we hold about ourselves that keep us stuck in repetitive patterns, ones that do not always lead to happiness, which are the patterns of asmita, or "I-ness." When asmita is loosened a little, our personal narrative begins to expand from one of "I-ness" to one of "we-ness," a feeling that we are interconnected. Our circle widens to embrace others and the world we live in as our extended body. As we become more established in a feeling of togetherness, of interconnectedness, the sense of "I" falls away, and the sense of "we" becomes strengthened. Eventually, even the idea of "we" falls away, for that is a boundary that separates, too. When the boundary of "we" dissolves, all that is left is being, existence. This is the mystical journey, a journey that is not bound by religion or owned by anyone, even the yogis. It is the state of being that describes what unity consciousness is. It is a journey that is there for anyone who feels called to investigate the three most important questions we can ask of ourselves:

Who am I?

What am I doing here?

What should I do next?

These are questions for all of us. And I join you in asking them of myself.

AFTERWORD

THE TEACHINGS OF YOGA and meditation have been passed down from generation to generation, in the guru-shishya, or teacher-student relationship, for many thousands of years. Like many other things, the form of yoga has changed to lesser and greater degrees over the centuries. Yoga one thousand years ago in all likelihood looked different than it does today. However, the essential underlying purpose of yoga has not changed from the earliest times until now, and that is as a way in which we seek to know who we truly are, through quieting the steady stream of thoughts, ideas, information, and images that fill our mind every day.

Krishnamacharya was a massive influencer in the world of yoga. Four of his disciples went on to become incredible influencers in their own right, and are responsible for at least half of the yoga that is practiced around the entire world today—and I say that without exaggeration. In 2018 there were an estimated 36 million people in America practicing some form of yoga, so worldwide, the number could be at least three to four times that. These four teachers include B.K.S. Iyengar, Krishnamacharya's nephew, who was a revolutionary figure in yoga. His approach was to meld anatomical exactitude of the postures with a fierce concentration-based practice. Indra Devi learned from Krishnamacharya in the 1930s; she was the darling of 1940s and '50s Hollywood, and was a precursor to the well-being lifestyle industry that later blossomed in California and influenced much of the American yoga and fitness crazes (see Stefanie Syman's book, *The Subtle Body*, for more on that). T.K.V. Desikachar started to study yoga seriously in the 1960s, and popularized his father's teachings on every continent. Krishnamacharya's earliest disciple was Sri K. Pattabhi Jois, who in 1937 organized Krishnamacharya's teachings into a systematic form, which was his own unique presentation of Ashtanga Yoga.

So we have Iyengar Yoga, practiced by literally millions of people, and popularized by people such as Yehudi Menuhin; Indra Devi's generalized influence on an entire movement of well-being that mushroomed out from Hollywood, California, and included famous actors, among them Gloria Swanson and Marilyn Monroe; and Desikachar's many students who went on to form the International Association of Yoga Therapists and other important institutions. Ashtanga Yoga has inspired at least two songs and one movie by Madonna and is practiced by other celebrities who have been vocal about their devotion to yoga, such as Gwyneth Paltrow, Willem Dafoe, Russell Brand, and Mike D of the Beastie Boys. It is taught in favelas in Rio, in the slums of Kenya, to victims of genocide in Rwanda, and in prisons in America, and has inspired yoga programs to reduce gun violence, help recovering addicts, and soothe traumatized victims of human trafficking in India. In short, yoga has gone everywhere, and much of its spread is because of these four teachers, who all share a common teacher: Sri Krishnamacharya.

The knowledge in the guru-shishya (teacher-student) tradition is tested knowledge, similar to a science experiment that has been proven to be replicable, and its results accepted in the larger scientific field. Yoga and meditative practices that are passed down contain knowledge based on experience. It's not because these practices are older that we say they are better—like the saying "Old is gold," for example; we don't say that the telegraph was better than a mobile phone is today because it came first.

A living teacher is absolutely necessary because we can only get so far on our own steam, and at a certain point we need to accept a guide who has gone to the depths of knowledge already and can lead us down paths that we have not yet seen, who can guide us when we struggle and encourage us when we become disillusioned. This acceptance of a teacher is called devotion, or *bhakti* in Sanskrit. The idea of devotion basically means to have faith or conviction in something other than the infallibility of your own abilities; it suggests that there is something deeper that powers us, and that sometimes we plain need help. Surrender does not mean giving up everything, including your own sense of agency; it means giving up the idea that you are the *only* source of agency.

A guru is a vessel who carries knowledge. *Gu* means "remover," and *ru* means "darkness." The darkness refers to seekers not knowing who they truly are, or what their purpose is. Darkness covers the inner light of knowledge, and yoga is a practice that can help remove darkness, just as turning on a light bulb removes the darkness from a room. The guru is not the light bulb; the guru is the one who can teach you how to turn on the switch, and in the end, you are the one who has to pull it. The *Katha Upanishad* explains that in the end it is the inner pull of the student that is the driving force behind transformation:

> *Nāyamātmā pravacanena labhyo na medhayā*
> *na bahunā śrutena |*
> *Yamevaisha vrnute tena labhya tasyaisha ātmā*
> *vrnute tanūgm svām ||*

||||||

> This Self cannot be known through much study, nor
> through the intellect, nor through much hearing
> (of teachings). It can be known by the Self alone that
> the aspirant seeks to know; the Self alone reveals its
> own nature to the seeker who is seeking it.

The Upanishads say that you don't have to constantly look outside yourself, for other people's knowledge, to know God or to know your true self; look straight to God to know God, and look straight toward your self to know your self. Truth reveals itself from our own desire to know, and not from any outside source. This is the ultimate teaching in self-reliance, but a self-reliance based on knowing who we truly are, and one that is fueled by humility, devotion, practice, and, most important, love. The information contained in this book is a glimpse of the fantastically intricate ways that yoga affects our body, nervous system, emotions, mind, and heart. They are topics that I find fascinating, and they inspire me to practice. And in the end, that is the essential thing: yoga is a practice, and if we do even just a little bit every day, the effects are compounded over time and carry over into the rest of our lives. If anything, I hope this book helps to inspire you to practice, too. If it does, then I consider it a success.

PRACTICE A

{ RESONANCE BREATHING }

Resonance breathing is essentially a reset button for your nervous system. Adults normally breathe at a rate of fifteen to eighteen breaths per minute, which ensures an adequate supply of oxygen to meet the body's energetic needs. Effective gas exchange also ensures that the blood pH is maintained at a suitable level, so that carbon dioxide can be removed from the body. Anxiety and stress often lead to slightly elevated rates of respiration, such as breathing rates of twenty or even up to twenty-five breaths per minute. When the breath speeds up but there's not a demand for faster breathing, a signal is sent to the brain that something is not right. This can lead to an activation of the fight-or-flight response of the sympathetic nervous system, and cause elevated blood pressure, inflammation, or other imbalances (including reinforcing the cycle of anxiety). When doing resonance breathing, we consciously slow our breath down to a cycle that brings the sympathetic and parasympathetic nervous systems into equilibrium, and thereby restore balance. This rate is generally five to seven breaths per minute, with the inhalation and exhalation being completely equal, or the exhalation being just ever so slightly longer than the inhalation. This is a breathing cycle that Tibetan monks and yogis naturally fall into when they meditate.

The term *resonance* refers to when two or more things or systems come into harmony with each other. As we discussed in the chapter on breath, our inhalations and exhalations are ruled by the sympathetic and parasympathetic nervous systems, and they are never completely in balance. One will always be slightly more dominant than the other, sometimes significantly so, such as when we are sleeping and the breath slows down, or when we are in a state of high stress and the breath speeds up. Resonance breathing is the only time when the sympathetic and parasympathetic nervous systems are at a total equilibrium due to the conscious regulation of the breath. Primary among the benefits is the balancing of the baroreflex, which involves the nerves that wrap around the carotid artery and control and monitor blood pressure. While doing paced breathing, we are consciously bringing our changing breathing patterns into an even rhythm in concert with the amount of time that it takes the baroreceptors to send messages to the heart as they monitor blood pressure. There is about a five-second delay between the pressure sensed in the carotid artery and the signals sent to the heart. We are also pacing the breathing with the fluidity of the vagal brake, and at the same time slowing the brain-wave pattern down to meditative frequencies. By virtue of slow exhalations, we tone the vagus nerve and stimulate the vagal brake, which is responsible for slowing the heart rate as we exhale. When this occurs, it is a sign of good cardiovascular health. When the heart does not respond to our breathing rate, it is an indication that something is off.

To recap from chapter 11, a toned vagus nerve is associated with:

* Reduction of inflammation
* Improvement in immune system function
* Reduction of anxiety and depression
* Improvement in the vagal braking mechanism
* Improvements with digestion

When the vagal tone is low, there may be cardiovascular problems, inflammation problems, and mood dysregulation.

Vagal tone is measured through heart rate variability (HRV), which is the beat-to-beat difference in our heart rate, and is directly correlated to our physiological and emotional health. The vagus nerve is responsible for controlling HRV. Dr. Richard Gevirtz, who specializes in HRV training through paced breathing—which is essentially the same as resonance breathing—has listed scientifically proven benefits of a slow breathing cycle. They are compelling and reason enough for anyone to try it. A slow breathing cycle:

* Tones the vagus nerve.
* Shuts off the stress response.
* Activates the relaxation response.
* Protects from pain in muscles.

* Reduces abdominal pain, especially in high performers.

* Supports healthy digestion.

* Lowers blood pressure.

Appreciation, love, and gratitude augment HRV, while anger, stress, and anxiety block its rhythm.

About two to three years ago I started practicing this type of breathing, and was amazed at the results that I saw and how quickly I saw them. For example, not only was I sleeping better, but for some reason I found that when I got in bed at night, I could access my parasympathetic nervous system's feeling of quiet and go right to sleep. I was learning to activate the vagal brake at will. My pranayama practice improved, because I could feel that my nervous system was becoming receptive and pliable from the resonance, and that breath holds were effortless and calm.

The method of practicing resonance breathing is quite simple. All you need to do is gradually slow your breathing rate down to five to seven breaths per minute, which means you will be breathing in any of the following ratios:

1. Inhaling for four seconds, exhaling for six seconds (six breaths per minute)

2. Inhaling for five seconds, exhaling for five seconds (six breaths per minute)

3. Inhaling for five seconds, exhaling for seven seconds (five breaths per minute)

4. Inhaling for six seconds, exhaling for six seconds (five breaths per minute)

5. Inhaling for four seconds, exhaling for four seconds (just under seven breaths per minute)

The breathing itself does not need to be deep; it just needs to be smooth and a little bit longer than a normal breath. If you try to breathe too deeply it may cause tension, or you might feel lightheaded. The first few times you begin to consciously change your breathing pattern, it might feel a little strange; that's why it is important not to overdo it, and start with just a minute or two, and gradually increase from there. It is most effective when it is done every day. It can be done lying down, with your legs up a wall, in a chair, or sitting in a meditative position, so choose any position that is comfortable for you. If you like, you can download the Breathing App that I developed with Deepak Chopra and Sergey Varichev, which has several different breathing cues and a timer. It will show a countdown clock if you need to know how much time you have left in your session. The timer starts at one minute and goes up to thirty minutes. If you only have time to practice for one or two minutes, that is fine—perhaps in between subway stops, or taking the elevator up to a job interview.

When you start the breathing practice, you can keep your hands on your belly, if that is comfortable for you. You should feel the belly move outward a little as you inhale, and in as you exhale. This is a good thing to get a sense for, because it will help get you in touch with the natural movement of the diaphragm, which moves down on the inhalation, pushing the belly outward, and up on the exhalation, so the belly moves back in.

Some people feel they can control the length of the breath better by slightly tightening the glottis muscle in the throat, which is what you do when you whisper or fog a mirror. This narrows the aperture where the breath flows, thus making it easier to regulate a smooth flow. However, it is not necessary, if you feel you don't need to do that.

Here are a few tips to keep in mind:

* It doesn't have to be "right"—there is no right breathing, since we breathe all the time. There is just some breathing that is more effective for gas exchange, and more effective for reducing levels of stress. Resonance is one of those types of breathing.

* In the beginning, don't try to do every breath. Just work on elongating the exhalation along with the cues. After that becomes comfortable, you can try elongating your inhalations to match the cues. Then, when you are ready, you can string the breaths together. Whenever you feel any stress building up (which happens on occasion), pause, go back to your regular breathing rhythm, and then when you are ready, try again. It sometimes takes a few minutes to settle in.

* Please try not to be self-critical or self-judgmental when you do the breathing. We are not looking to become expert breathers; we are just trying to balance our nervous system with it.

* Let your belly gradually relax, and try not to breathe too much in your chest. Some sensation of breath flow will move upward into the chest, which is of course fine because that is where your lungs are. However, when given the cue to inhale, it is a natural response to take a deep breath into the chest. We are trying to do the opposite, which means that when given the cue to inhale, we relax and let the breath drop down toward the bottom part of the lungs, rather than pull the breath into the upper part of the lungs. The effects will be deeper and more profound within your nervous system.

※ Give yourself a few minutes to settle in. The calm feeling that accompanies resonance breathing happens after the nervous system gets accustomed to the ratio, so go slow and give yourself time. If you try to rush it, it will take longer to work.

※ Breathe with a feeling of kindness, love, or compassion for yourself or for the process of breathing. This has a wonderful effect on the mind, heart, and vagus nerve.

When you are finished with your session, sit quietly for a moment or two. The quiet space you have created within yourself is a good place for reflection. Ask yourself the following questions, without looking for answers, just letting the questions ripple in your field of consciousness like the ripples made in water when a small pebble is dropped into a lake:

Who am I?

What is my deepest desire?

What am I grateful for?

If answers come, it is fine, but they don't have to. Feel as though you are absorbing the effects of the breathing into your cellular body, and into your nervous system's memory. Soak in the peaceful feeling like a sponge.

Bring this equilibrium of awareness with you into your day. If there is a project, a goal you have, or a problem you need to work out, engage it with your sense of "I am." Remember that each person you are interacting with, either physically or mentally, also operates with this very same sense of "I am." The consciousness that lies underneath the feeling of "I am" is the same in you as it is in that person, as it is in me and everyone else. Resonance breathing gives you back your sense of you. It puts you back in the driver's seat, when the increasing load of the world so often takes over. With resonance breathing we spend a few minutes each day living and breathing in a perfect balance between sympathetic and para-sympathetic, in a neutral zone of mental calm and equilibrium, of conscious peace.

When you are done with your session, and have soaked in the benefits for a minute or so, you can get up and go about your day.

{ UNILATERAL, OR SINGLE-NOSTRIL, BREATHING }

This is a very simple and effective breathing practice. It is classified as *nadi shodana*, which means "nerve purification." It is not considered a pranayama practice since there is no breath retention (kumbhaka), which is the defining characteristic of pranayama. It is useful for anyone who experiences stress, or feels that they need to refresh their brain at any point during the day. I have taught this to yoga students from the ages of eleven to eighty. It is considered safe and useful for anyone who does not have any serious nasal obstructions, or a disease that would prevent them from practicing elongated breathing. Most people find that just a few rounds of single-nostril breathing is refreshing, calming, and focusing for the brain and mind. Studies have shown unilateral nostril breathing to reduce anxiety and improve verbal and spatial cognition. Its method of practice goes like this:

* Sit comfortably in a cross-legged position, in a chair, or on your bed if you have a condition that is keeping you in bed.

* Place your right thumb against the indentation on the right side of your nose where your nostril is meeting the cartilage of the nose.

* Breathe smoothly and comfortably through the left nostril three times.

* If your left nostril is very blocked, only lightly press the right nostril, but leave a little opening for some air to pass through.

* After you have taken three breaths, release your hand down to your lap or knee and breathe through both nostrils two or three times, sensing any difference in the airflow.

* Then change nostril sides by placing your right ring finger on your left nostril, and breath three times smoothly and freely through your right nostril.

* After you are done, rest your hand on your knee, and breathe a few times through both nostrils.

This practice can be done in the morning or evening. It should be done at least an hour after eating any food, and thirty minutes after any liquids. You can also do it in the afternoon when you need a replenishment break, perhaps instead of something that has sugar or caffeine. As you get comfortable with the number of rounds, you can increase them, but let it come naturally. More is not the goal. When your arm feels tired, your face feels flushed, or you feel the inklings of any tension, you have done enough. It is also a good practice for before or after meditation, to balance the hemispheres of the brain.

LOVING-KINDNESS MEDITATION

In 2013, Dr. Bethany Kok published a study that investigated the connection between vagal tone, positive emotions, and physical health. Improvements in vagal tone have been shown in studies to be associated with positive emotions improving physical health, and also with physical health improving positive emotions. What is not known is what the underlying mechanism is that ties positive emotions to vagal tone and physical health. Kok and her colleagues hypothesized that there is an "upward-spiral dynamic" that reinforces the emotional-physical health connection and is linked to how people perceive their social connections. The intervention they used to test this was a loving-kindness meditation, with a meditation class once a week and daily self-directed practice at home for six weeks. The results of the study were overwhelmingly positive. The participants who reported the greater increases in positive emotions were the ones who had been randomly assigned to the meditation group. They exhibited larger increases in both social connections and vagal tone. "Results suggest that positive emotions, positive social connections, and physical health influence one another in a self-sustaining upward-spiral dynamic." Even more promising, it "also raises the possibility that changes in habitual emotions drive changes in vagal tone, and thereby constitute one pathway through which emotional health influences physical health. We propose that people's ability to translate their own positive emotions into positive social connections with others may be one of the keys to solving this mystery."[1]

One of the earliest versions of the loving-kindness meditation can be found in the *Yoga Sutras*, chapter 1, verse 33. Here we find described a practice that brings brightness and clarity to the mind through focusing on our relationships with other people, our social connections. Everything up until verse 33 has involved us primarily working with our own minds. Interestingly enough, Patanjali is careful to point out that when it comes to spiritual evolution, it's not all just about us, but also about how we relate to others. The verse says:

Maitri karuna mudita upeksha sukha dukha punya
apunya vishayanam bhavana tah citta prasadanam ||

||||||

Purification of mind comes from having an attitude of
friendliness toward those who are happy; compassion
toward those who are suffering; sympathetic joy toward
those who are virtuous; and equanimity of mind
toward those who are un-virtuous. 1.33

These four attitudes can be purposefully adopted by us and directed to the different types of relationships we have. They can help shape a positive perception of our social connections, even toward people who annoy us, make us angry, or cause us to feel justifiably, righteously indignant. Sometimes happy people make us feel annoyed, suffering people make us want to turn away, those who are doing wonderful things in the world make us jealous, and some who are behaving horribly make us outraged. This verse suggests a simple remedy: For those who are happy, be friendly toward them. For those who are suffering, feel their pain as your own, expressing compassion, but do not try to fix them—we can't remove the pain that others suffer from, but we can sit with them through it. For those are doing wondrous things in the world, have sympathetic joy and feel their successes as your own. Perhaps the most difficult are those who are, well, difficult. Toward those people, Patanjali suggests having equanimity of mind. Try not to be ruffled by them. In the end, it is we who will suffer because of our state of mind, so we might as well avoid suffering by keeping our minds cool and calm. As Kok says in the opening line of her study, "People who experience warmer, more upbeat emotions live longer and healthier lives." Who wouldn't want that?

The Buddhists call these four qualities the *brahma viharas*, and they are the foundational practices of loving-kindness. They are also called the "four immeasurables," because through their practice the mind is said to become immeasurable, like space. These practices are pre-Buddhistic but were adapted by the Buddha into his canon and have become one of the preeminent practices done by Buddhists today. In fact, loving-kindness meditation is much more popular these days among Buddhists than among yogis. My wife, Jocelyne, has been practicing loving-kindness meditation for the past fourteen years in line with the Buddhist teachings (and practicing yoga for the past thirty years), and what follows are her instructions on how to practice loving-kindness meditation.

Loving-kindness meditation is comprised of the repetition of some simple phrases. With each repetition, we are expressing intention, planting the seeds of loving wishes over and over in our heart. With a loving heart as a background, all that we attempt, all that we encounter, will open and flow easily.

We begin by directing the phrases toward ourselves. You can recite inwardly or out loud these traditional phrases directed to your own well-being:

May I be safe.

May I be happy.

May I be healthy.

May I be peaceful.

We begin with ourselves because without loving ourselves it is impossible to love others. We repeat these phrases over and over, letting the feelings permeate our body and mind. Once you are comfortable with the phrases, you can expand them:

May I be safe and protected from inner and outer harm.

May I be peaceful and happy.

May my body be strong and healthy, and heal itself over and over.

May I take care of myself with ease and joy.

The experience of loving-kindness may not happen immediately. It can seem a little uncomfortable at first to sit and repeat the phrases, especially when we don't understand what we are doing. It also takes effort and concentration to repeat the phrases. Sometimes you might get sleepy or bored while repeating the phrases; if this happens, don't judge yourself. Boredom happens sometimes. Try to notice if you feel shut down. You do not need to manufacture some type of artificial loving energy. It is the loving-kindness itself that will eventually break the barrier. When we don't seem to connect, we need to back off and trust the process instead of feeling like a failure. Bring gentleness and patience to the process. Tune in to your own goodness, or a beautiful quality of someone else that touches your heart. That is the connection—it can be as simple as an appreciation of something good.

We are all very different in the way that we find our natural entry point to the connection to the quality of the heart, to the tranquil energy of the heart. The instructions might feel too much sometimes, and they might feel like they don't resonate with you.

Some teachers start with the forgiveness practice first, to purify the heart—and only after that do they teach the loving-kindness practice. The forgiveness practice goes like this:

> *To whom I have hurt knowingly or unknowingly, I ask*
> *forgiveness.*
>
> *To those who have hurt me knowingly or unknowingly,*
> *I forgive.*
>
> *I forgive myself for hurting myself knowingly or*
> *unknowingly.*

You can repeat each phrase four times, and repeat the whole process three times.

This is a good way to free the heart, free the mind, and make it so there is no room for hatred.

It is important within these practices that you set your intention, so gentleness and patience come into play as you repeat the words. The connection will come easier, and then the heart can stretch. Once you have connected, you will get a felt sense of ease. You won't feel separate, and you will feel in touch with some kind of true bond. After, you need to maintain the connection and generate this sense of tenderness toward others.

You can also check: What does the energy of kindness do? If you receive kindness from someone, let's say unconditionally—it can be a very simple gesture, such as someone opening the door for you—you may feel touched and appreciative. Inside you may feel a shift when it happens. Then you might want to open the door for someone else, too. Kindness spreads itself; we don't actually spread it!

If you like, you can try a weeklong experiment with loving-kindness.

DAY 1

Start with some breathing techniques, such as resonance or single-nostril breathing, and make sure you are in a position where you can be comfortable for fifteen minutes. Do the forgiveness practice for ten to fifteen minutes: repeat the phrases slowly, bring in mind images and circumstances, and try to infuse the words of forgiveness as you go along.

Keep it simple. Use your breath to anchor the forgiveness phrases.

DAY 2

For ten to fifteen minutes, repeat the loving-kindness phrases to yourself in a slow manner. You can reread the material on kindness, connecting with your own goodness as you breathe in and out. Breathe in long but comfortable breaths, and as you breathe out, visualize and repeat the phrases, nourishing your self with both the breath and the phrases. Try to notice any effects your practice has on your emotions and mental state during the day.

Do you feel, perhaps, more ease, calmness, or happiness during the day?

Sometimes the practice can bring up a sense of hurt, disconnection, unworthiness, or fear. At this time, we can also bring in some extra phrases, such as: "May I accept myself for the way I am," or other phrases that you feel recall good qualities or actions that you are proud of. Allow yourself to embody the positive energy in a gentle, calming way, to establish balance and a positive attitude toward yourself. Using the breath and words, try to get a felt sense of a place or a time (for example, when you were a child) when you felt strong and happy. Take in those feelings, and soak them in. Feel yourself soften. Feel yourself soften in your body. Create within yourself a safe space.

DAYS 3 AND 4

Sit comfortably, as is becoming your habit.

Take a few breaths and tune in to your intention that you are now going to do loving-kindness meditation.

Relax your whole body.

Focus on your heart center, having a felt sense of the body sitting and breathing.

Bring to mind a benefactor, a teacher or friend, someone who is dear to you and whom you trust fully, and who has supported you in the past. Visualize them smiling, or sense inwardly their voice, their energy, their warmth, and an internal connection.

Imagine the benefactor wishing the phrases to you, and you receiving the phrases. Receive the safety, protection, and security from that person. Be the receiver of happiness, peace, kindness, wishes of good health. Allow yourself to receive unconditionally.

Breathing in and out, take your time. Then you may offer the phrases back to your benefactor, and feel that they are receiving them in turn. See how you feel when you offer the phrases to the benefactor. Notice if there are any changes in your body, or how you feel energetically. If you get lost in your thoughts, begin again—it's normal. You can use your imagination extending well wishes, and you can imagine the benefactor receiving them. See how that feels inside you.

When the heart is tender, it feels good. When we are greedy, angry, or jealous, it really does not feel good. When we are patient, our heart is lighter. When we give, it feels good. When we let go, it feels good. This practice is also called purification of the heart. When you begin again, bring the image of the person first, then the felt sense of them, then your breath, and then the words. The words should feel as if they are carried by the breath.

DAYS 5 AND 6

You may do the same as previously if you feel that you have established a good momentum with the benefactor, yourself, and the phrases. If the phrases don't resonate with you, you can drop them and experiment with your own phrases. See how your mind works. Some people do not visualize easily but are better with words. Some are better with sensing the presence of the people to whom the phrases are directed, but not with the words. Be flexible. For some, just repeating the words by themselves is enough.

Loving-kindness is like building up a fire. We keep adding log after log. Now we will bring the neutral person category. The connection with the neutral person is fairly easy. It can be a teller at the bank, a police officer, a post office person, or someone you see regularly in a store or on the street but you don't know. Just as you wish for yourself to be at ease, happy, healthy, and balanced, you can share these same wishes with the neutral person. That will be your connection, and your thread.

You can visualize them, and send them the phrases of well wishes. When we feel happy, we give unconditionally; we want sincerely for everyone to be at peace, ease, and balance. The muscle of the heart gets stronger.

DAY 7

Now for the meditation on the "difficult person." It is not always easy when we get to the difficult person, sometimes called "the enemy." Whom do we throw out of our hearts, and why? We think if we let them in, they might take advantage of us, or we feel vulnerable. The Indian saint Neem Karoli Baba said, "Despite how much you dislike somebody, never let them out of your heart."

When we close our heart, we suffer, and we are the ones who get caught in the snare of sorrow. Loving-kindness works on the relationship without the other person knowing. So we work on our own, and it changes the dynamic of the relationship and leaves room for openness. We don't do it to change the other person; we do it to transform our heart and to release the pain we are carrying. Forgiveness is a good way to release the knot.

When you sit to meditate, make sure you keep the eyes and jaw soft, as well as the shoulders, as you repeat the phrases. Be aware to keep the body relaxed, checking in with the comfort of your body from time to time. Let your breath carry the energy, and let the breath release the energy.

There are other techniques and suggestions that some teachers offer as ways to work with a difficult person. You can choose which of the following resonate most with you:

* One way to deal with the difficult person is to feel as though you can create a bigger container for your own self to be established in, before sending loving-kindness

to the difficult person. Create a field of balance and equanimity before you start. In your body you can feel such things as balance or the steadiness of sitting like a mountain. Some teachers suggest imagining that the difficult person is as vulnerable as an infant, and then you can allow yourself to feel their fragility rather than their harshness. The Buddha said, "Hatred can never cease by hatred. Hatred can only cease by love. This is an eternal law."

* Some suggest remembering that your difficult person is suffering, too.

* Try to build a more neutral energy, an evenness of the body, and mind and heart equanimity.

* Remember that as someone is a difficult person for you, you might also be a difficult person for them.

* Recognize the truth that hurts people and the truth that heals people.

When we feel a bit stuck in opposition to the difficult person, we need to step out. We may think about the person's loved ones or people who love them. It can bring a thread of respect and some extra space. In this way you can see the person's humanity, and see if you can ride with it.

Throughout the day, try to infuse kindness into your daily activities. For example:

* While brushing your teeth

* While you get dressed

* When you go to bed

Try to be open and curious. You might even change some of your habits, or do things differently, like how you get out of bed, eat, talk, or think. Bring some awareness to where you could fit more self-care into your life. Notice when you resist, when you judge yourself, if you are challenged with some aspect of yourself that you struggle with.

Notice the people around you, and wish the same goodness for them as you wish for yourself. Think of your family, friends, and pets, and wish them the same. It's a well-wishing habit, and when you do it in a meaningful way, you can feel the effect it has on your feelings of kindness, and your feelings toward yourself, all living beings, nature, vibrating out into the universe.

BODY SCAN

The most vivid memory I have of my first yoga class was experiencing the deep calm of "final relaxation," the progressive relaxation that is typically taught at the end of a yoga class. I was fourteen or fifteen at the time, at a summer camp, and remember going into a state that was devoid of any thoughts, dreams, or images, and I remember feeling clearly that this is what death must feel like. Absolute nothingness, with no identity, time, space, or fear. I had wanted to continue trying yoga on my own at home, but was afraid that I would not know how to wake myself up from the final relaxation. Perhaps the teacher knew a special trick to wake us, and what if I went into that deep state, and my mother found me, but didn't know how to wake me up? Would I be stuck there, in nothingness? It was a genuine fear, so I didn't do yoga again until I was about eighteen or nineteen, when I was able to find a teacher in New York.

The practice of progressive relaxation that is taught in yoga classes is not found in the yoga texts, but something called *savasana* is. In the *Hatha Pradipika*, chapter 1, verse 34, savasana is taught as a posture that removes fatigue and gives rest to the mind, and is done by "laying your body on the earth like a dead body which has fallen face up."[1] However, no instruction on visualization or muscle relaxation is given. Pattabhi Jois taught savasana as the imitation of a body in rigor mortis, the stage several hours after death when the body becomes rigid. In his approach, the student holds the body completely rigid, and if you try to pick them up by the head, or by their feet, the body will not bend at all and come up in one straight line. It is an advanced asana.

The savasana that I learned in my first yoga class, and that is done in yoga classes all over the world, is actually a combination of savasana (lying on the ground resting), and progressive muscle relaxation, which was created in 1908 by a doctor of internal medicine, psychiatry, and physiology from Chicago named Edmund Jacobsen. Jacobsen was also the inventor of biofeedback, which uses the paced breathing of resonance breathing. Through measuring muscle tone and nerve impulses, Jacobsen was able to prove a connection between muscular tension and different disorders of the body and mind. He found that the shortening of muscle fibers due to tension decreased their tone, which decreased certain activities of the central nervous system. Through relaxing tension in the muscles, the aggravating pull on the nervous system would also be released, resulting in relief from a variety of diseases. He may well have popularized the word *relax* to be synonymous with releasing tension from both mind and musculature, in the same way the Canadian endocrinologist Hans Selye redefined the word *stress* to be synonymous with the stimulus of environmental demands. Indeed, almost everyone who does yoga has at one point or another said, "I'm here for the final relaxation!"

There are several different ways of practicing deep relaxation, which can also be done as a "body scan," a popular Buddhist meditation practice. The effects can be quite profound, anything from an enhanced sense of relaxation to transcendent experiences of consciousness. Jacobsen's findings that the release of muscular tension leads to the release of the nervous system holds quite true, and by consciously relaxing each and every body part and organ, and even the act of thinking itself, we can enter into deep states of quiet. Below are two different ways you can try to practice it.

METHOD ONE

Lie either on your back, side, or abdomen—whichever is most comfortable for you. If you are on your back, and feel any back discomfort, you can bend your knees so that your feet are flat on the ground, about hip width apart. Follow along with the script below, internally repeating each phrase, and allowing your awareness to move through each part of your body. In Jacobsen's method, one would squeeze each limb tightly before relaxing it, but that it not necessary to do. It is not comfortable for everyone to follow that method. The word *relaxing* is only one option of a guiding word; you can replace it with *softening* or *resting* if another word resonates with you. For some people, the word *relax* itself can be stressful.

I am relaxing my feet, I am relaxing my feet; my feet are relaxed.

I am relaxing my ankles, I am relaxing my ankles; my ankles are relaxed.

I am relaxing my legs, I am relaxing my legs; my legs are relaxed.

I am relaxing my hips, I am relaxing my hips; my hips are relaxed.

I am relaxing my hands, I am relaxing my hands; my hands are relaxed.

I am relaxing my arms, I am relaxing my arms; my arms are relaxed.

I am relaxing my shoulders, I am relaxing my shoulders; my shoulders are relaxed.

I am relaxing my abdomen, I am relaxing my abdomen; my abdomen is relaxed.

I am relaxing my chest, I am relaxing my chest; my chest is relaxed.

I am relaxing my back, I am relaxing my back; my back is relaxed.

I am relaxing my neck, I am relaxing my neck; my neck is relaxed.

I am relaxing the back of my head, I am relaxing the back of my head; the back of my head is relaxed.

I am relaxing behind my ears, I am relaxing behind my ears; behind my ears are relaxed.

I am relaxing my scalp, I am relaxing my scalp; my scalp is relaxed.

I am relaxing my forehead, I am relaxing my forehead; my forehead is relaxed.

I am relaxing my eyes, my nose, and my mouth, I am

relaxing my eyes, my nose, and my mouth; my eyes, my nose, and my mouth are relaxed.

I am relaxing my cheeks and my chin, I am relaxing my cheeks and my chin; my cheeks and my chin are relaxed.

I am relaxing my heart and my lungs, I am relaxing my heart and my lungs; my heart and my lungs are relaxed.

I am relaxing my stomach, I am relaxing my stomach; my stomach is relaxed.

I am relaxing my liver, I am relaxing my liver; my liver is relaxed.

I am relaxing my spleen, I am relaxing my spleen; my spleen is relaxed.

I am relaxing my pancreas, I am relaxing my pancreas; my pancreas is relaxed.

I am relaxing my intestines, I am relaxing my intestines; my intestines are relaxed.

I am relaxing my kidneys and my adrenal glands, I am relaxing my kidneys and my adrenal glands; my kidneys and my adrenal glands are relaxed.

I am relaxing my pineal gland, my pituitary gland, my thyroid gland, and my sex glands; I am relaxing my pineal gland, my pituitary gland, my thyroid gland, and my sex glands; my pineal gland, my pituitary gland, my thyroid gland, and my sex glands are relaxed.

I am relaxing my brain, I am relaxing my brain; my brain is relaxed.

I am relaxing my nervous system, I am relaxing my nervous
    system; my nervous system is relaxed.

I am relaxing my vagus nerve, I am relaxing my vagus
    nerve; my vagus nerve is relaxed.

I am relaxing my breath, I am relaxing my breath; my
    breathing is relaxed.

I am relaxing my thinking, I am relaxing my thinking; my
    thoughts are relaxed.

I am relaxing my inner being, I am relaxing my inner
    being; my inner being is relaxed.

For just these next few minutes, I'll let myself deeply relax,
    resting in a calm, quiet, neutral space. Anything I need
    to think about, I can think about after I rest. For now,
    I am letting go into rest.

To come out from your resting state, slowly lengthen your
breathing, gently bringing movement back into your body through
your breathing. Stretch the arms and legs in any way you would
like to, as if you were getting up in the morning. When you are
ready, you can get up and go about your day, carrying your calm
within you.

METHOD TWO

This method of relaxation is based on the space meditation found
in Insight Meditation. It can be done sitting up or lying down. The
idea behind this technique is that tension, clinging to our ideas of
righteousness, and stress come from holding ourselves tightly in our
body or mind. Tight spaces are the opposite of open spaces. Space,
in the meditative traditions, is said to be a place of non-attachment,

relaxation, expansion, and love. The instructions below are from my wife, Jocelyne.

The way this practice is done is by enquiry. We will investigate the different spaces of the body by imagining or becoming aware of the space between two different points, or regions, of the body.

To begin:

* Place your hands on your belly, and slowly breathe in to a count of five.

* Pause.

* Slowly breathe out to a count of six.

* Do this three times to slow down the nervous system.

As you sit comfortably, visualize the space between the eyes. Can you feel the space between your eyes?

Pause, and then imagine the space between the ears, the inside of the nose, the cavern of the mouth, pausing between each. Can you feel the space within each part of the face?

Then, imagine the space of the throat. Can you feel the space in your throat?

Imagine the space between the shoulders, between the arms and hands, right and left. Can you feel the space between the shoulders, and between each arm and hand?

Imagine the space of the lungs, and the space between the breastbone and the spine. Can you feel the space between your breastbone and spine?

Imagine the space in the heart. Can you feel space in your heart?

Imagine the space between the navel and the spine. Can you feel the space between your navel and spine?

Imagine the space between the hips. Can you feel the space between your hips and pelvic bones?

Imagine the volume of the legs and space of the feet.

Imagine the space of the room: floor, ceiling, windows, doors, and any other details that you like. Feel the space above and below the body, and all around you. Tune into your heart with a sense of ease, and see if you can find a thread of tenderness, and let it gently extend infinitely beyond the most faraway star.

Take a few breaths, and let the whole experience unfold.

REMINDERS

You want to let the breath take you to the different spaces, or to open the doors to the different spaces moment to moment. Let the momentum flow with ease, not trying to manipulate or get anything out of it. Be as intimate as possible with breath and space, and relaxed. Do not worry or think that it should be pleasant, or that it is unpleasant.

Some days the energy will be stronger, some other days lower. You can bring more curiosity to your mind to awaken your interest and energy. You can imagine the breath like a flashlight, lighting the space on the way.

You might notice your breath going deeper or shorter, your heart beating faster, louder, or softer sometimes. You might feel underlying emotions, like sadness, anger, or happiness, which will trigger the movement of the breath or attention to shift. Thoughts can be tempting to follow at times. Keep it simple. Keep the momentum of your awareness steady, and stay connected. It will build energy on its own.

As you deepen in the feeling of space, can you become aware of your thoughts and feelings, passing by like clouds in the sky?

To finish, allow yourself to come to a place of silence and space, a sense of moment-to-moment presence that cannot be disturbed.

Let your heart breathe with silence and space.

As you get more comfortable with imagining or feeling space within your body, you can expand to move through the parts of the body with more detail. Here is a map that you can follow if you wish to go a little deeper:

THE HEAD

* Space between the eyes
* Space between the nasal passages
* Space that fills the tongue, upper palate, teeth, gums, and soft palate
* Space that fills the lips
* The whole mouth at once
* The entire head and face

THE THROAT AND NECK

* Space in the throat and neck

UPPER BODY

* Space of the region of the shoulders
* Space of the upper arms
* Space of the lower arms
* Space inside the thumb and the index finger, including flesh and bone

* Space between middle finger, ring finger, and pinky

* Space between the palm of the hand and the top of the hand

* Simultaneously imagine the space inside the fingers, hands, arms, and shoulders.

* Can you imagine the space of the lungs and the heart?

* Can you imagine the space between the breastbone and the spine?

* Can you imagine the space in the entire chest?

LOWER TORSO

* Space between the navel and the spine

* The distance between the sides of the waist

* Space between the navel and the base of the spine

* Imagine the space of the whole belly and the whole lower torso at the same time.

LOWER BODY

* Space between the hip joints and the knees

* Space of the regions of the upper legs

* Space of the regions of the lower legs

* Space that fills the big toe, and all the toes

* Space between the top of the foot and the sole of the foot

* Space of both feet

Recall all of the inner space that you have been opening up to with your breath, from top to bottom, for a few breaths, in and out. Wash away all the sensations of the solidity of the body, almost as a melting of the boundaries of the body, which becomes an extension of the universe. Let your breath be like the wind that blows as an extension of the wind in the world. No inside, and no outside. When you are ready, you can bring yourself slowly out from the meditation, and gently go about your day.

NOTES

1. WHAT IS YOGA?

1. Sebastien Manrique, *Travels of Fray Sebastien Manrique 1629–1643*, vol. 1 (New York: Routledge, 2016), https://books.google.com/books?id=pAckDwAAQBAJ&printsec=frontcover&dq=Travels+of+Fray+Sebastian&hl=en&sa=X&ved=0ahUKEwjTuJCn8ZPbAhVrp1kKHca1D3YQ6AEIMzAC#v=onepage&q=Travels%20of%20Fray%20Sebastian&f=false.

2. See Ananda Bhattacharya's fascinating *A History of the Dasnami Naga Sannyasis.*

3. David N. Lorenzen, "Warrior Ascetics in Indian History," *Journal of the American Oriental Society* 98, no. 1 (Jan.–Mar. 1978): 61–75, https://www.jstor.org/stable/600151?origin=crossref&seq=1#page_scan_tab_contents.

4. Krishnamacharya's 1934 publication, *Yoga Makaranda,* written at the behest of the Maharaja of Mysore, was largely aimed at reinvigorating

the practice of yoga, which was disappearing from India as a result of Western influence. "Our own youngsters who have the necessary skill and intelligence to compete with foreigners will, I am sure, resurrect and uplift our culture."

5. Swami Sivananda, of Rishikesh, made two all-India tours to disseminate knowledge of yoga and spirituality. In 1950 he traveled to North and South India, as well as Sri Lanka (then Ceylon), seeing millions of people along the way. His notes are chronicled in *Sivananda's Lectures During All India and Ceylon Tour, 1950,* edited by Swami Venkatesananda (Rishikesh, India: Divine Life Publications, 2009).

6. Annie Gowen, "India's New Prime Minister, Narenda Modi, Aims to Rebrand and Promote Yoga in India," *Washington Post*, December 2, 2014: "Shripad Yesso Naik, India's new yoga minister, dreams of a day when sun salutations and downward-facing dog pose will be as popular in their homeland as they are around the world," https://www .washingtonpost.com/world/asia_pacific/indias-new-prime-minister -narendra-modi-wants-to-rebrand-and-promote-yoga-in-india/2014/12 /02/7c5291de-7006-11e4-a2c2-478179fd0489_story.html?utm_term= .0e4cd7edc2a3.

7. A quick Internet search will show lists of the many organizations that provide these services. The David Lynch Foundation, for example, specializes in teaching Transcendental Meditation to veterans, women who have been rescued from trafficking, and children in schools, to name just one organization.

8. For example, the Encinitas, California, yoga trial in 2015 sought to distinguish between yoga as a social-emotional wellness practice and a religious practice: *Steven Sedlock et al. v. Timothy Baird et al.*, Court of Appeals, Fourth Appellate, Division One, State of California, D064888, April 3, 2015, https://cases.justia.com/california/court-of-appeal/2015 -d064888.pdf?ts=1428084026.

9. See the *Yoga Sutras* 2.31: "However, the restraints become a great vow when they become universal, and are unrestricted in regards to the class of birth, country, time, or concept of duty (of any person)" (*Jātideśakālasamayānavichhinnāh sārvabhaumā mahāvratam*).

10. Patañjali, *Yoga Philosophy of Patañjali: Containing His Yoga Aphorisms with Vyāsa's Commentary in Sanskrit and a Translation with Annotations Including Many Suggestions for the Practice of Yoga*, annotated by Swami Hariharānanda Āraṇya (Albany: State University of New York Press, 1983), 3.

11. The six are the *Brahma Sutras*; *Yoga Sutras*; *Purva Mimamsa Sutras*; *Vaisheshika* (or *Kanada*) *Sutras*; *Sankhya Karika*; and *Nyaya Sutras*.

12. Yogaschittavritti nirodhah, *Yoga Sutras* 1.2.

13. Patañjali, *Yoga Philosophy of Patañjali*, 1.

14. Ibid., 8.

15. Sri K. Pattabhi Jois, *Yoga Mala: The Original Teaching of Ashtanga Yoga Master Sri K. Pattabhi Jois* (New York: North Point Press, 1999), 4.

16. Vivekakhyātir aviplavā hānopāh, *Yoga Sutras* 2.26.

17. Jois, *Yoga Mala*, 4–5.

3. THE PRACTICE OF POSTURES

1. Billye Anne Cheatum and Allison A. Hammond, *Physical Activities for Improving Children's Learning and Behavior: A Guide to Sensory Motor Development* (Champaign, Ill.: Human Kinetics, 2000), 34–35.

2. Dennis S. Charney and Steven M. Southwick, *Resilience: The Science of Mastering Life's Greatest Challenges* (New York: Cambridge University Press, 2012), 35–36.

3. Rami Sivan, *Theory and Practice of Hindu Ritual,* vol. 1 (Sri Matham), http://www.srimatham.com/uploads/5/5/4/9/5549439/hindu_ritual_vol _1.pdf, 12–13.

4. With the commentary Jyotsnā of Brahmānanda and English translation, *The Hathayogapradīpikā of Svātmārāma* (Chennai, India: The Adyar Library and Research Centre, 1972), 11.

5. Lothar Schäfer, *Infinite Potential: What Quantum Physics Reveals About How We Should Live* (New York: Deepak Chopra Books, 2013), 8.

6. Sivan, *Theory and Practice of Hindu Ritual*, 13.

7. P. T. Katzmarzyk, Timothy S. Church, Cora Lynn Craig, and Claire Bouchard, "Sitting Time and Mortality from All Causes, Cardiovascular, Disease, and Cancer," *Medicine and Science in Sports and Exercise* 41, no. 5 (May 2009): 998–1005.

8. J. S. Jaiswal and L. L. Williams, "A Glimpse of Ayurveda: The Forgotten History and Principles of Indian Traditional Medicine," *Journal of Traditional and Complementary Medicine* 7, no. 1 (2015): "Digestive fire is important in controlling the normal microflora, proper digestive functions and provision of energy to the entire body. Any disturbances in its balance, creates discomfort to the gastro-intestinal tract and results in pathological complications like ulcers, diarrhea and constipation," https://www.researchgate.net/publication/305448610_A_glimpse_of

Ayurveda-_The_forgotten_history_and_principles_of_Indian _traditional_medicine.

9. *The Hathayogapradīpikā of Svātmārāma*, 11.

10. Patañjali, *Yoga Philosophy of Patañjali*, 19.

11. Rick Hanson, *Hardwiring Happiness: The New Brain Science of Contentment, Calm, and Confidence* (New York: Harmony Books, 2013), 20.

12. Perhaps one day, an intrepid yogi or researcher will track down a copy of it. If you want to look, I'd start in Calcutta. Of course, it is almost certain that Krishnamacharya had other yogic influences as well, as can be seen from the bibliographies of his books, *Yoga Makaranda* and *Yogasanagalu*. In the bibliography of *Yoga Makaranda* the *Yoga Korunta* is not listed, but in *Yogasanagalu* it is.

13. This is in line with Pattabhi Jois's comment to me and his grandson Sharath, in conversation in 2006, that Krishnamacharya taught the students at the Palace Yoga Shala in Mysore in the 1930s asanas one after the other with no particular separation of categories. When Pattabhi Jois was given the post of creating a yoga department at the Sanskrit College in Mysore, he created a four-year syllabus of asanas, pranayama, philosophy, and Sanskrit grammar that formed the basis for the yoga system he taught for the rest of his life. He said to Sharath and me that he presented his groupings of asanas to Krishnamacharya, who gave his approval. This was in 1937. It was not until four years later, in 1941, that Krishnamacharya came out with his second book, *Yogasanagalu*, in which he included groupings of asanas similar to Pattabhi Jois's syllabus. These groupings, called Primary, Middle, and Proficient or Advanced, are more or less the same as Pattabhi Jois's groupings. The list of asanas in *Yogasanagalu* can be found at Anthony Grim Hall, "Krishnamacharya's Yogasanagalu (1941) (translation project)," *Krishnamacharya's Mysore Yoga . . . at Home*, n.d. (blog post), http://grimmly2007.blogspot.com/p /yogasanagalu-translation-project.html.

14. Michael Joyner and Darren P. Casey, "Regulation of Increased Blood Flow (Hyperemia) to Muscles During Exercise: A Hierarchy of Competing Physiological Needs," *Physiological Reviews* 95, no. 2 (April 2015): 549–601; and Walter F. Boron and Emile L. Boulpaep, *Medical Physiology: A Cellular and Molecular Approach*, 2nd ed. (Philadelphia: Saunders Elsevier, 2012), 467.

4. THE SEAT OF AWARENESS

1. Daniel J. Siegel, *Brainstorm: The Power and Purpose of the Teenage Brain* (New York: Jeremy P. Tarcher/Penguin Group, 2013), 47–48.

2. These nine drishtis are listed in one of the *Yoga Korunta* verses that Pattabhi Jois learned orally from Krishnamacharya: *Nāsāgre netrayor-madhye nābicakras tathaiva ca hastāgre pādayoragre pārśvayor ubhayor api aṅguṣṭhāgre urdhva-dṛṣṭiḥ navadṛṣṭi-prakīrtitāḥ.*

3. Maurizio Corbetta et al., "A Common Network of Functional Areas for Attention and Eye Movements," *Neuron* 21, no. 4 (Oct. 1998): 761–73.

4. Daniel Kahneman, *Thinking, Fast and Slow* (New York: Farrar, Straus and Giroux, 2011). See pp. 32–35: "Much like the electricity meter outside your house or apartment, the pupils offer an index of the current rate at which mental energy is used."

5. Michal T. Kucewicz et al., "Pupil Size Reflects Successful Encoding and Recall in Memory in Humans," *Scientific Reports* 8 (2018), https://www.ncbi.nlm.nih.gov/pubmed/29563536.

6. John J. Ratey, with Eric Hagerman, *Spark: The Revolutionary New Science of Exercise and the Brain* (New York: Little, Brown, 2008), 41.

7. Sat Bir Singh Khalsa's book *The Principles and Practice of Yoga in Health Care* is an invaluable and comprehensive collection of yoga studies on everything from mental health conditions to cancer.

8. Marshall Hagins and Andrew Rundle, "Yoga Improves Academic Performance in Urban High School Students Compared to Physical Education: A Randomized Controlled Trial," *Mind, Brain, and Education* 10, no. 2 (May 2016): 105–16; M. Hagins, A. Rundle, N. S. Consedine, and S. B. Khalsa, "A Randomized Controlled Trial Comparing the Effects of Yoga with an Active Control on Ambulatory Blood Pressure in Individuals with Prehypertension and Stage 1 Hypertension," *Journal of Clinical Hypertension* 16, no. 1 (Jan. 4, 2014), PubMed PMID: 24387700; D. Wang and M. Hagins, "Perceived Benefits of Yoga Among Urban School Students: A Qualitative Analysis," *Evidence-Based Complementary and Alternative Medicine*, 2016, Article ID 8725654, http://dx.doi.org/10.1155/2016/8725654; M. Hagins, R. States, T. Selfe, and K. Innes, "Effectiveness of Yoga for Hypertension: Systematic Review and Meta-analysis," *Evidence-Based Complementary and Alternative Medicine* 2013, 2013:649836, doi: 10.1155/2013/649836; L. Daly, S. Haden, M. Hagins, N. Papouhis, and P. Ramirez, "Yoga and Emotional Regulation in High

School Students: A Randomized Controlled Trial," *Evidence-Based Complementary and Alternative Medicine* 2015, Article ID 794928, doi:10.1155/2015/794928; and S. Haden, L. Daly, and M. Hagins, "A Randomised Controlled Trial Comparing the Impact of Yoga and Physical Education on the Emotional and Behavioural Functioning of Middle School Children," *Focus on Alternative and Complementary Therapies* 19, no. 3 (Sept. 2014): 148–55.

9. Gudrun Bühnemann, *Eighty-Four Asanas in Yoga: A Survey of Traditions* (New Delhi: DK Printworld, 2007), 20–21: "This view of the subordinate position of asanas clearly differs from that of most modern Yoga schools." And p. 22: "Even so, modern Yoga schools regularly invoke the authority of YS despite the fact that there is little in their curricula that bears resemblance to the teachings in that text." Mark Singleton, *Yoga Body: The Origins of Modern Posture Practice* (London: Oxford University Press, 2010), 27: "In spite of the scarcity of information regarding asana in the sutras themselves and in the traditional commentaries, the text is routinely invoked as the source and authority of modern postural yoga practice."

10. T. S. Rukmani, trans., *Yogasūtrabhaāsyavivarana of Śankara: Vivarana Text with English Translation and Critical Notes Along with the Text and English Translation of Patanjali's Yogasūtras and Vyāsabhāsya,* vol. 1 (New Delhi, India: 2010), 369.

11. Estimation by head of research, Dr. M. L. Gharote, James Russell, "Yoga Korunta: Unearthing an Ashtanga Legend," James Russell Yoga, Nov. 11, 2015 (blog post), http://www.jamesrussellyoga.co.uk/blog-james -russell_files/Yoga%20Korunta%20-%20unearthing%20an%20Ashtanga %20legend.html.

12. Karl Baier, Philipp A. Maas, and Karin Preisendanz, eds., "The Proliferation of Āsanas in Late Medieval Yoga Texts," *Yoga in Transformation: Historical and Contemporary Perspectives on a Global Phenomenon* (Vienna: Vienna University Press, 2018), 131.

13. T.K.V. Desikachar and Kausthub Desikachar, trans., *Adi Sankara's Yoga Taravali* (Chennai, India: Krishnamacharya Yoga Mandiram, 2003): "*Sahasrashah santu hathesu kumbhah sambhavyate kevala kumbha eva | kumbhottame yatra tu rechapurau pranasya na prakrtvai-krtakhyau ||* The school of Hatha yoga talks about thousands of kinds of pranayama. Among them, Kevala Kumbhaka is the most respected. In this state of Kevala Kumbhaka, there is no movement of breath outward or inward."

5. WHERE IS MY MIND?

1. Neeta Mehta, "Mind-Body Dualism: A Critique from a Health Perspective," *Mens Sana Monograph* 9, no. 1 (Jan.–Dec. 2011): 202–209: "This method involved the breaking up of a problem into pieces and rearranging them in a logical order. Under the spell of the 'scientific revolution' that positivism brought in, disciplines like physics, chemistry and astronomy not only flourished but also came to define exact science. The success of the scientific method reinforced Descartes' philosophy and methodology further and contributed to the dogma of scientism (Klein and Lyytinen, 1985), the belief that scientific method was the only legitimate path to knowledge."

2. Gilbert Ryle, "Descartes' Myth," in *The Concept of Mind* (London: Hutchinson, 1949), 13: "Such in outline is the official theory. I shall often speak of it, with deliberate abusiveness, as 'the dogma of the Ghost in the Machine.' I hope to prove that it is entirely false, and false not in detail but in principle. It is not merely an assemblage of particular mistakes. It is one big mistake and a mistake of a special kind. It is, namely, a category mistake."

3. This paragraph is largely from information gathered from Esther M. Sternberg, *The Balance Within: The Science of Connecting Health and Emotions* (New York: W. H. Freeman, 2001).

4. *Mundaka Upanishad* III.ii.4.

5. The Wheel of Awareness can be found online on Dr. Siegel's website, https://www.drdansiegel.com.

6. WHO AM I?

1. James Baldwin, *The Price of the Ticket* (New York: St. Martin's Press, 1958), 244.

2. Possibly from *Namarupa Magazine*, but I misplaced the reference in my notebook where I wrote this quote down.

3. Elizabeth Avedon, *An Interview with Francesco Clemente by Rainer Crone and Georgia Marsh* (New York: Vintage Books, 1987), 19–20.

7. THE FIRST TWO LIMBS

1. *Ahimsa pratishtayam tat sannidau vairatyagaha*, *Yoga Sutras* 2.35.

2. *Satyam bruyat priyam bruyan na bruyat satyam apriyam | priyam cha nanrtam bruyadesha sanatanah || Manusmriti* 4.138.

3. *Satya pratishtayam kriyaphalashrayatvam*, *Yoga Sutras* 2.36.

4. *Asteya pratishtayam sarvaratnopasthanam*, *Yoga Sutras* 2.37.

5. *Brahmacharya pratishtayama virya labhah*, *Yoga Sutras* 2.38.

6. *Aparigrahastairyam janmakathantasambodha*, *Yoga Sutras* 2.39.

7. *Shauchatsvangajugupsa parairsamsargah*, *Yoga Sutras* 2.40. The basis for this is the yogi's realization that the body can never become fully clean, as day after day it gets dirty again, and therefore the body will always be a source of uncleanliness.

8. *Santoshadanuttamasukhalabha*, *Yoga Sutras* 2.43.

9. *Kayendriyasiddhirashuddhiksayattapasah*, *Yoga Sutras* 2.43.

10. *Svadhyayadishtadevatasamprayogah*, *Yoga Sutras* 2.44.

11. *Samadhisiddhirishvarapranidhanat*, *Yoga Sutras* 2.45.

8. INTERNAL ENERGY

1. Bessel van der Kolk, *The Body Keeps the Score: Brain, Mind, and Body in the Healing of Trauma* (New York: Penguin Books, 2015), 249.

2. OpenStax College, *Anatomy and Physiology* (Houston: Rice University, 2013), 1061.

3. Eric P. Widmaier, Hershel Raff, and Kevin T. Strang, *Vander's Human Physiology: The Mechanisms of Body Function*, 10th ed. (New York: McGraw-Hill Higher Education, 2006), 605.

4. Ibid., 439.

5. Ibid., 508.

6. Stephen W. Porges, *The Polyvagal Theory: Neurophysiological Foundations of Emotions, Attachment, Communication, and Self-Regulation* (New York: W. W. Norton, 2011), 288.

7. T. Pramanik, B. Pudasaini, and R. Prajapati, "Immediate Effect of a Slow Pace Breathing Exercise Bhramari Pranayama on Blood Pressure and Heart Rate," *Nepal Medical College Journal* 12, no. 3 (2010): 154–57.

9. BREATH AS SPIRIT

1. This passage from the *Chandogya Upanishad* is fairly long, and stretches from 7.1.2 to 7.15.1. The above is my truncated adaptation of the conversation.

2. *Chandogya Upanishad* 5.1.7–5.1.12.

3. *Taittiriya Upanishad* 2.4.

4. *Chale vate chalam chittam nischale nischalam bhavet | Yogi sthanutvamapnoti tato vayum nirodhayet ‖ Hathayogapradipika* 2.2.

5. M. C. Melnychuk, P. M. Dockree, R. G. O'Connell, P. R. Mur-

phy, J. H. Balsters, and I. H. Robertson, "Coupling of Respiration and Attention via the Locus Coeruleus: Effects of Meditation and Pranayama," *Psychophysiology* (April 22, 2018), https://doi.org/10.1111/psyp.13091.

6. Ido Amihai and Maria Kozhevnikov, "The Influence of Buddhist Meditation Traditions on the Autonomic System and Attention," *BioMed Research International* (2015), Article ID 731579, http://dx.doi.org/10.1155/2015/731579.

10. TIPS ON PRACTICE

1. *Bhagavad Gita* 6.17.

2. Marilia Carabotti, Annunziata Scirocco, Maria Antonietta Maselli, and Carola Severi, "The Gut-Brain Axis: Interactions Between Enteric Microbiota, Central and Enteric Nervous Systems," *Annals of Gastroenterology* 28, no. 2 (April–June 2015): 203–209.

3. Deepak Chopra and Rudolph E. Tanzi, *Super Genes: Unlock the Astonishing Power of Your DNA for Optimum Health and Well-Being* (New York: Harmony Books, 2015), 87–88.

4. Q. Feng, W. D. Chen, and Y. D. Wang, "Gut Microbiota: An Integral Moderator in Health and Disease," *Front Microbiology* (Feb. 2018), https://www.ncbi.nlm.nih.gov/pubmed/29515527.

5. R. Sender, S. Fuchs, and R. Milo, "Are We Really Vastly Outnumbered? Revisiting the Ratio of Bacterial to Host Cells in Humans," *Cell* 164, no. 3 (Jan. 2016), 337–40. From the abstract: "It is often presented as common knowledge that, in the human body, bacteria outnumber human cells by a ratio of at least 10:1. Revisiting the question, we find that the ratio is much closer to 1:1." Also: American Microbiome Institute, "How Many Bacteria vs Human Cells Are in the Body?" Jan. 20, 2016 (blog post), http://www.microbiomeinstitute.org/blog/2016/1/20/how-many-bacterial-vs-human-cells-are-in-the-body.

6. L. A. David et al., "Diet Rapidly and Reproducibly Alters the Human Gut Microbiome," *Nature* 505 (Jan. 23, 2014): 559–63, https://www.ncbi.nlm.nih.gov/pubmed/24336217/.

7. Gil Sharon et al., "The Central Nervous System and the Gut Microbiome," *Cell* 167, no. 4 (Nov. 2016): 915–32.

8. N. A. Jessen et al., "The Glymphatic System: A Beginner's Guide," *Neurochemical Research* 40, no. 12 (Dec. 2015): 2583–99, https://www.ncbi.nlm.nih.gov/pubmed/25947369.

9. *Svapna nidra jnana alambana va, Yoga Sutras* 1.38.

11. THE NERVOUS SYSTEM, EAST AND WEST

1. *Shariram adyam khalu dharma sadhanam, Dharma Shastras.*

2. Shirley Telles et al., "Hemispheric Specific EEG Related to Alternate Nostril Yoga Breathing," *BMC Research Notes* 10, no. 1 (2017), https://bmcresnotes.biomedcentral.com/articles/10.1186/s13104-017-2625-6.

3. Tania Lombrozo, "The Truth About the Left Brain/Right Brain Relationship," Dec. 2, 2013, https://www.npr.org/sections/13.7/2013/12/02/248089436/the-truth-about-the-left-brain-right-brain-relationship.

4. S. A. Jelle and D. S. Shannohoff-Khalsa, "The Effects of Unilateral Forced Nostril Breathing on Cognitive Performance," *International Journal of Neuroscience* 57, nos. 3–4 (Nov. 1993): 73, https://www.ncbi.nlm.nih.gov/pubmed/8132419.

5. NYU Langone Health, "Researchers Find New 'Organ' Missed by Gold Standard Methods for Visualizing Anatomy and Disease," news release, March 27, 2018, https://nyulangone.org/press-releases/researchers-find-new-organ-missed-by-gold-standard-methods-for-visualizing-anatomy-disease.

6. The Indian system of Marma points, which are similar to the acupuncture points but which I know less about, should also be mentioned here.

7. There are other communication systems, such as the immune system, that operate independently of the nervous system.

8. *Katha Upanishad* 2.3.11.

9. Carabotti et al., "The Gut-Brain Axis."

10. Bruce H. Lipton, *The Biology of Belief,* 10th Anniversary Edition (Carlsbad, Calif.: Hay House, 2016), 10.

11. See the documentary *Slomo* by Josh Izenberg for an amazing demonstration of this: Josh Izenberg, director, *Slomo, New York Times* Op-Docs, Mar. 31, 2014, https://www.nytimes.com/2014/04/01/opinion/slomo.html.

12. R. Buckminster Fuller, with Jerome Agel and Quentin Fiore, *I Seem to Be a Verb* (New York: Bantam Books, 1970).

13. Alexandru Barboi, "Sympathy, Sympathetic: Evolution of a Concept and Relevance to Current Understanding of Autonomic Disorders," *Neurology* 80, no. 7, supplement (March 21, 2103), http://n.neurology.org/content/80/7_Supplement/S57.005.short.

14. Telles et al., "Hemisphere Specific EEG," 306.

15. Bethany Kok et al., "How Positive Emotions Build Physical Health: Perceived Positive Social Connections Account for the Upward

Spiral Between Positive Emotions and Vagal Tone," *Psychological Science* 24, no. 7 (May 6, 2013): 1123–32, http://journals.sagepub.com/doi/full/10.1177/0956797612470827.

16. Scott E. Krahl, "Vagus Nerve Stimulation for Epilepsy: A Review of the Periphreal Mechanisms," *Surgical Neurology International*, 2012, https://www.ncbi.nlm.nih.gov/pmc/articles/PMC3400480/.

17. Porges, *The Polyvagal Theory*, 264.

18. Ibid., 14, 151.

19. Ibid., 14.

20. Stephen W. Porges, "The Polyvagal Perspective," *Biological Psychology* 74, no. 2 (March 2007): 116–43.

21. Divya Krishnakumar, Michael R. Hamblin, and Shanmugamurthy Lakshmanan, "Meditation and Yoga Can Modulate Brain Mechanisms That Affect Behavior and Anxiety: A Modern Scientific Perspective," *Ancient Science* 2, no. 1 (April 2015): 13–19, https://www.ncbi.nlm.nih.gov/pmc/articles/PMC4769029/. There are several important similarities and differences between hormones and neurotransmitters. Both effect changes in the body and behavior. The activities of the nervous system, for the most part, depend on the release of neurotransmitters to carry messages and instructions to the body. The endocrine system depends on hormones. Hormones, such as adrenaline and cortisol, are produced by the glands that make up the endocrine system. They are released directly into the bloodstream, and effect behavioral and bodily changes slowly. Neurotransmitters, such as acetycholine and dopamine, are chemical messages released by the firing of neurons. They effect changes in the body quickly. Ninety percent of the neurons that release serotonin, an important mood regulator, are found in the gut.

22. Amy F. T. Arnsten, Murray A. Raskind, Fletcher B. Taylor, and Daniel F. Connor, "The Effects of Stress Exposure on Prefrontal Cortex: Translating Basic Research into Successful Treatments for Post-Traumatic Stress Disorder," *Neurobiology of Stress* 1 (Jan. 2015): 89–99, https://www.sciencedirect.com/science/article/pii/S2352289514000101.

23. Matthew A. Pimental, Ming G. Chai, Caroline P. Le, Steven W. Cole, and Erica K. Sloan, "Sympathetic Nervous System Regulation of Metastasis," in *Metastatic Cancer: Clinical and Biologocal Perspectives*, ed. Rahul Jandial (Austin, Tex.: Landes Bioscience, 2013).

24. Patañjali, *Yoga Philosophy of Patañjali*, 225.

25. Porges, *The Polyvagal Theory*, 140.

26. Ibid., 94–95.

27. Center for Compassion and Altruism Research and Education, Stanford University, *Science of Compassion 2014: The Psychophysiology of Compassion* (video), n.d., http://ccare.stanford.edu/videos/science-of -compassion-2014-the-psychophysiology-of-compassion/.

28. A. G. Mohan, trans., with Ganesh Mohan, *Yoga Yajnavalkya* (Madras, India: Ganesh & Co.), verse 12.8.

29. Vasant G. Rele, *The Mysterious Kundalini: The Physical Basis of the "Kundalini Yoga"* (Bombay, India: D. B. Taraporevala Sons & Co., 1927), 65.

30. T.K.V. Desikachar and Kausthub Desikachar, trans., *Yoga Taravali* (Chennai, India: Krishnamacharya Yoga Mandiram, 2003), verses 13–15.

31. J. C. Smith, H. H. Ellenberger, K. Ballanyi, D. W. Richter, and J. L. Feldman, "Pre-Bötzinger Complex: A Brainstem Region That May Generate Respiratory Rhythm in Mammals," *Science* 254, no. 5032 (Nov. 1991): 726–29, doi:10.1126/science.1683005, PMC 3209964, PMID 1683005.

32. W. C. Bushell, "Longevity: Potential Life Span and Health Span Enhancement Through Practice of the Basic Yoga Meditation Regimen," *Annals of the New York Academy of Sciences* 1172, no. 1 (August 2009): 20–27.

33. K. B. Hengen, T. M. Gomez, K. M. Stang, S. M. Johnson, and M. Behan, "Changes in Ventral Respiration Column GABAaR ϵ-and δ-Subunits During Hibernation Mediate Resistance to Depression by EtOH and Pentobarbital," *American Journal of Physiology* 300, no. 2 (Feb. 2011), https://www.ncbi.nlm.nih.gov/pmc/articles/PMC3043800/.

PRACTICE C: LOVING-KINDNESS MEDITATION

1. Bethany E. Kok, Kimberly A. Coffey, Michael A. Cohn, Lahnna I. Catalino, Tanya Vacharkulksemsuk, Sara B. Algoe, Mary Brantley, and Barbara L. Fredrickson, "How Positive Emotions Build Physical Health," *Psychological Science* 24, no. 7 (2013), 1123–32.

PRACTICE D: BODY SCAN

1. Kausthub Desikachar, trans., *The Hathayogapradīpikā: Jyotsnāyutā* (Chennai, India: Media Garuda, Krishnamacharya Healing & Yoga Foundation, 2016), 34.

BIBLIOGRAPHY

Avedon, Elizabeth. *An Interview with Francesco Clemente by Rainer Crone and Georgia Marsh*. New York: Vintage Books, 1987.

Baldwin, James. *The Price of the Ticket*. New York: St. Martin's Press, 1958.

Bühnemann, Gudrun. *Eighty-Four Asanas in Yoga: A Survey of Traditions*. New Delhi, India: DK Printworld (P) Ltd, 2007.

Bhattacharyya, Ananda, ed. *A History of the Dasnami Naga Sannyasis*. New York: Routledge, 2018. https://books.google.com/books?id =O1JPDwAAQBAJ&pg=PT7&lpg=PT7&dq=naga+sannyasis+and +east+india+company&source=bl&ots=-uaCyEAX5f&sig =stkQHlkzfdRYjH_df2oNOgh7B8s&hl=en&sa=X&ved=0ahUKE wi25Ny4tYLbAhXDslkKHXJxDu4Q6AEIOTAB#v=onepage&q =naga%20sannyasis%20and%20east%20india%20company&f=false.

Chāndogya Upaniṣad. With the commentary of Śaṅkarācārya. Translated by Swāmī Gambhīrānanda. Calcutta, India: Advaita Ashrama, 1983.

Charney, Dennis S., and Steven M. Southwick. *Resilience: The Science of Mastering Life's Greatest Challenges.* New York: Cambridge University Press, 2012.

Cheatum, Billye Anne, and Allison A. Hammond. *Physical Activities for Improving Children's Learning and Behavior: A Guide to Sensory Motor Development.* Champaign, Ill.: Human Kinetics, 2000.

Chopra, Deepak, and Rudolph E. Tanzi. *Super Genes: Unlock the Astonishing Power of Your DNA for Optimum Health and Well-Being.* New York: Harmony Books, 2015.

Desikachar, Kausthub, trans. *The Hathayogapradīpikā: Jyotsnāyutā.* Chennai, India: Media Garuda, Krishnamacharya Healing & Yoga Foundation, 2016.

Desikachar, T.K.V., and Kausthub Desikachar, trans. *Yoga Taravali.* Chennai, India: Krishnamacharya Yoga Mandiram, 2003.

Eight Upanisads, vol. 2. With the commentary of Śaṅkarācārya. Translated by Swāmī Gambhīrānanda. Calcutta, India: Advaita Ashrama, 1992.

Fuller, R. Buckminster, with Jerome Agel and Quentin Fiore. *I Seem to Be a Verb.* New York: Bantam Books, 1970.

Hanson, Rick. *Hardwiring Happiness: The New Brain Science of Contentment, Calm, and Confidence.* New York: Harmony Books, 2013.

The Hathayogapradīpikā of Svātmārāma. With the Commentary Jyotsnā of Brahmānanda and English Translation. Chennai, India: The Adyar Library and Research Centre, 1972.

Jois, Sri K. Pattabhi. *Yoga Mala: The Original Teaching of Ashtanga Yoga Master Sri K. Pattabhi Jois.* New York: North Point Press, 1999.

Kahneman, Daniel. *Thinking, Fast and Slow.* New York: Farrar, Straus and Giroux, 2011.

Khalsa, Sat Bir Singh, Lorenzo Cohen, Timothy McCall, and Shirley Telles, eds. *The Principles and Practice of Yoga in Health Care.* Edinburgh: Handspring Press, 2016.

Mohan, A. G., trans., with Ganesh Mohan. *Yoga Yajnavalkya.* Madras, India: Ganesh & Co., 2013.

OpenStax College. *Anatomy and Physiology.* Houston, Tex.: Rice University, 2013.

Patañjali. *Yoga Philosophy of Patañjali: Containing His Yoga Aphorisms with Vyāsa's Commentary in Sanskrit and a Translation with Annota-*

tions Including Many Suggestions for the Practice of Yoga. Annotated by Swami Hariharānanda Āraṇya. Albany: State University of New York Press, 1983.

Porges, Stephen W. *The Polyvagal Theory: Neurophysiological Foundations of Emotions, Attachment, Communication, and Self-Regulation.* New York: W. W. Norton, 2011.

Ratey, John J., with Eric Hagerman. *Spark: The Revolutionary New Science of Exercise and the Brain.* New York: Little, Brown, 2008.

Sarvānanda, Swāmī. *Taittirīyopaniṣad.* Madras, India: Sri Ramakrishna Math, 1965.

Schäfer, Lothar. *Infinite Potential: What Quantum Physics Reveals About How We Should Live.* New York: Deepak Chopra Books, 2013.

Siegel, Daniel J. *Brainstorm: The Power and Purpose of the Teenage Brain.* New York: Jeremy P. Tarcher/Penguin Group, 2013.

Siegel, Daniel J. *Mind: A Journey to the Heart of Being Human.* New York: W. W. Norton, 2017.

van der Kolk, Bessel. *The Body Keeps the Score: Brain, Mind, and Body in the Healing of Trauma.* New York: Penguin Books, 2015.

van Lysebeth, André. *Pranayama: The Yoga of Breathing.* London: Unwin Paperbacks, 1983.

Widmaier, Eric P., Hershel Raff, and Kevin Strang. *Vander's Human Physiology: The Mechanisms of Body Function*, 10th ed. New York: McGraw Hill Higher Education, 2006.

Yogasūtrabhaāsyavivarana of Śankara, vol. 1. Vivarana text with English translation and critical notes along with the text and English translation of Patanjali's Yogasūtra's and Vyaāsabhāsya. Translated by T.S. Rukmani. New Delhi, India: Munshiram Manoharlal Publishers, 2010.

ACKNOWLEDGMENTS

To my mom and dad, who always said when I was in high school, "We don't care how well you do as long as you try your hardest." Well, I never did very well, and to be honest, I also didn't try very hard. I needed to find something that I truly loved to try hard for it, and that was yoga. I'm glad I found it early. My ninth-grade science teacher once wrote in a report card that until I got a haircut and stopped wearing ripped jeans, I wouldn't ever do well in science. I think that he'd be pleasantly surprised to see that science has now become a huge part of my life. But I'm not thanking him for that, I'm thanking my parents for laughing that off. Our family is a big one, and I love them all. My two sisters, Kara and Amanda, are my rocks and always have been, and always will be. My half-sister Nina is a whole in our hearts. My other half-but-wholes are Nick and Rebecca, and the steps who were never steps away are Kathy, James, John, and Mary. I come somewhere in the middle. To my stepmother, Sallie, "Thank you, but I don't eat chicken" will always make all of us laugh. And last of the family is my stepfather, Jimmy, who passed away in 1993. I flew home from India just in time to see him one last time; we all miss him still, immensely.

I am deeply indebted to the following people for the love, support, knowledge, and guidance they have given me over the years:

To Sri K. Pattabhi Jois and R. Sharath Jois, for welcoming and guiding me and so many others onto the ancient path of yoga;

To Jeff Seroy, for championing this book through to completion and for our continued twenty-year friendship that started with *Yoga Mala*;

To Deepak Chopra, my friend, collaborator, and fellow seeker, who opened up a world of knowledge through his teaching and introduced me to many of the people who inspired my research, answered my questions, and gave me guidance, including Murali Doraiswamy, William Bushell, Subhash Kok, and Neil Theise;

To my oldest yoga friend, one of my first teachers, and my publishing partner, Robert Moses, who is a constant source of feedback, guidance, and inspiration;

To Marshall Hagins, collaborator, friend, and research partner, whose careful corrections and articulation of scientific facts in this book have helped to ensure that I was able to say the things I wanted to in the most accurate way possible;

To Samuel Collombet for the many suggestions he made on the science aspects of this book. His suggestion to reorder the presentation of chapter 11 helped not only make the section much more readable, but also helped clarify my thinking on it. I thank him also for his patient answering of all of my questions about science—questions that he answers even when I ask them while I'm giving a lecture;

To the many teachers who have hosted me in their yoga schools, making it possible for me to explore the material in this book: in alphabetical order, Jens Bache, Jenny Barrett-Bouwer, Dmitry Baryshnikov, Susanna Finocchi, Maarten van Huijstee, Deborah Ifill, Juha Javanainen, Jackie Kleefeld, Lisa Laler, Elena de Martin, Wessel Pater-

notte, Claudia Pradella, Petri Raisenen, Dr. Darshan Shah, and Priya
Shah;

To Alexandra Seidenstein for being my science tutor and answering all of my questions about everything from homeostasis to how cells work; Goran Bell for the "cellular death while sitting" research; James Bouwer for explaining deep aspects of the suprachiasmatic nucleus; Paul Dallaghan for turning me on to rhythm pattern generators; Rabbi Mendel Jacobsen for all of his instruction over the past year in Judaism and Kabbala; Dr. Kiran Bhat for the information he gave me about the internal anal sphincter and bradycardia; Sheetal Shah for her insightful suggestions; and Leeah Chu for her support during the writing process;

A special thanks to my cousin, endocrinologist Julia Chafkin, for introducing me to my vagus nerve hero, Stephen Porges;

To my friend and *Yoga Sutras* study partner Francesco Clemente, with whom I can speak mystic whenever I feel like it.

A massive thank-you to the scientists who have spearheaded all of the research that made a lot of this book possible, especially Sat Bir Khalsa, Stephen Porges, Shirley Telles, Bethany Kok, Rick Hanson, and the many, many others who have taken part in studies that I have cited or read.

A huge expression of gratitude for the music of Sigur Rós and Nick Cave and the Bad Seeds, whose music I had on repeat during the writing of this book, particularly Sigur Rós's *Ágaetis Byrjun*, *Kveikur*, and *Takk*, and Nick Cave's *Live at KCRW* and *Push the Sky Away*. Also, everything by David Bowie from 1971 to 1980.

Finally, there is no way that I can accomplish anything in my life without the love of my wife, Jocelyne Stern (the real yogi of our family), who supports and encourages me in every way possible, even when I probably don't deserve it, and our daughter, Lili, who is magnificent, kind and beautiful beyond measure, and smarter than the two of us combined.